PHOBIC GEOGRA[

T0231621

Joyce Davidson provides a powerful and highly o.
and at the same time helps all of us to think afresh about ordinary, everyday
experiences at the boundaries between our selves and our environments. By
exploring phobic experiences, Davidson sheds new light on what it means to be a
gendered, embodied and situated subject.

Liz Bondi, Professor of Social Geography, The University of Edinburgh

This book does a wonderful job in weaving together a sophisticated
theoretical framework with the voices and experiences of women. Each enriches
the other in ways that will surely make a creative contribution to the work of
a wide range of readers.

Dr Gillian Rose, Senior Lecturer in Geography, The Open University

For my mum and dad, and for Jim.

Phobic Geographies

The Phenomenology and Spatiality of Identity

JOYCE DAVIDSON

Queen's University, Kingston, Ontario

Routledge
Taylor & Francis Group

LONDON AND NEW YORK

First published 2003 by Ashgate Publishing

Published 2017 by Routledge
2 Park Square, Milton Park, Abingdon, Oxfordshire OX14 4RN
711 Third Avenue, New York, NY 10017, USA

First issued in paperback 2017

Routledge is an imprint of the Taylor & Francis Group, an informa business

Copyright © Joyce Davidson 2003

Joyce Davidson has asserted her right under the Copyright, Designs and Patents Act, 1988, to be identified as author of this work.

All rights reserved. No part of this book may be reprinted or reproduced or utilised in any form or by any electronic, mechanical, or other means, now known or hereafter invented, including photocopying and recording, or in any information storage or retrieval system, without permission in writing from the publishers.

Notice:
Product or corporate names may be trademarks or registered trademarks, and are used only for identification and explanation without intent to infringe.

British Library Cataloguing in Publication Data
Davidson, Joyce, 1971-
    Phobic geographies : the phenomenology and spatiality of
    identity
    1. Agoraphobia 2. Feminist geography
    I. Title
    304.2

Library of Congress Cataloging-in-Publication Data
Davidson, Joyce, 1971-
    Phobic geographies : the phenomenology and spatiality of identity / Joyce Davidson.
        p. cm.
    Includes bibliographical references and index.
    ISBN 0-7546-3244-X
    1. Agoraphobia. 2. Human geography. 3. Medical geography. 4. Women--Mental health.
    I. Title.

RC552.A44D386 2003
616.85'225--dc21
                                                                              2003042190

ISBN 13: 978-1-138-27792-2 (pbk)
ISBN 13: 978-0-7546-3244-3 (hbk)

Contents

Acknowledgements

Mick Smith has contributed to this project from a time before it was even conceived and his close involvement throughout has been of profound importance. He has taken part in discussions about all aspects of this book and the research on which it is based, from the pettiest methodological details to my most substantial conceptual concerns. Mick has suffered for my research like no one person should, and I could never thank him enough.

Liz Bondi has supported this project unfailingly since its inception, and her insight, sensitivity and hard work have been absolutely invaluable. Gillian Rose, too, became involved at a very early stage, and has remained committed to reading and advising in detail and depth throughout. Her support and enthusiasm, too, have been vital. Liz and Gillian were both involved in setting up the (Edinburgh) Feminist Geography Reading Group, whose important role I would also like to acknowledge. This inspirational group of women encouraged me in countless ways.

While the research for this book took place during my time at the University of Edinburgh, the book itself was largely written at Lancaster University, while I was employed as an NHS funded Post-Doctoral Research Fellow (award number RDO/35'12). I would like to thank each and every one of my colleagues at the Institute for Health Research, but especially Tony Gatrell and Christine Milligan for providing the space and encouragement to complete this project, and Vicki Bell for her excellent proof reading and indexing skills. Thanks too, to Amanda Bingley and Rosaleen Duffy for providing friendship and the space to sleep!

Several agoraphobia support organisations provided help and information for this study, including The National Phobics Society, No Panic, Pax and TOP, and especially StressWatch Scotland, WASP, and WLMI. The respondents who were interviewed for this project were all members of the latter three, and to these, individually and collectively, I owe an enormous debt of gratitude. The research was funded by the Economic and Social Research Council (award number R00429834370), and their support too is gratefully acknowledged.

I am very grateful to the following for granting permission to incorporate copyright material that has previously been published elsewhere. Blackwell Publishing for "" ... the world was getting smaller": women, agoraphobia and bodily boundaries', *Area*, 32, 1 (2000): 31-40, and for 'A phenomenology of fear: Merleau-Ponty and agoraphobic life-worlds', *Sociology of Health and Illness*, 22, 5 (2000): 640-660. Routledge for 'Fear and trembling in the mall: women, agoraphobia and body boundaries', in Isabel Dyck, Nancy Davis Lewis and Sara McLafferty (eds.) *Geographies of Women's Health* (2001): 213-230. Taylor and Francis Limited for 'Pregnant pauses: agoraphobic embodiment and the limits of (im)pregnability', *Gender, Place and Culture*, 8, 3 (2001): 283-297, and for '"Joking apart ... ": a "processual" approach to researching self-help groups', *Social and*

Cultural Geography, 2, 2 (2001): 163-183. Rowman and Littlefield Publishers for '"All in the mind?"': women, agoraphobia and the subject of self-help', in Liz Bondi, Hannah Avis, Ruth Bankey, Amanda Bingley, Joyce Davidson, Rosaleen Duffy, Victoria Ingrid Einagel, Anja-Maaike Green, Lynda Johnston, Susan Lilley, Carina Listerborn, Mona Marshy, Shonagh McEwan, Niamh O'Connor, Gillian Rose, Bella Vivat and Nichola Wood, *Subjectivities, Knowledges and Feminist Geographies: The Subjects and Ethics of Social Research* (2002): 15-33. Pion Ltd for '"Putting on a face": Sartre, Goffman and agoraphobic anxiety in social space', *Environment and Planning D: Society and Space*, 21, 1 (2003).

Finally, I would like to thank my family: my mum and dad for their constant love, support and friendship; wee Sam, Furd and Horus for being wonderful and wise; and Jim, the best big brother ever, who I'll never stop missing.

Introduction

Notes on Stories, Selves and Spaces

Robin Kearns has described one role of the geographer as a 'teller of tales' (1997: 269). In this book, I aim to take up this positioning in order to communicate something of the stories told by a largely neglected group – agoraphobic women – about their restricted and debilitated 'life-worlds' (Dyck 1995; Edwards and Uhlenhuth 1998). Agoraphobics tend to fear and avoid social space and women account for around 89 per cent of diagnosed agoraphobics (Clum and Knowles 1991; Bekker 1996), a proportion comparable to those suffering from anorexia nervosa (Battersby 1998: 46. See also Bordo 1993). But, while there has been considerable interest in anorexia by feminists and social scientists concerned to explain its gendered distribution, the same cannot be said for agoraphobia. And, although there has been no shortage of *clinical* research into agoraphobia and related anxiety disorders, these have as yet attracted relatively little attention outside of their medical and psychological contexts. This book constitutes the beginnings of an attempt to remedy these shortcomings.

My aims in this book are twofold; to provide non-clinical, specifically geographical insights into agoraphobia of relevance for its sufferers, and, complementarily, to expand human geography understandings of the relations between gender, embodiment, space and mental health, via a study of agoraphobia. It will be argued that, through its quest to fulfil these aims, the project reveals innovative potential in terms of both *therapy* and *theory*.

Therapeutically, I suggest that a study conducted from a feminist geographical perspective can improve existing models of agoraphobia, in ways that could have valuable implications for treatment. Existing treatment models tend to be based on clinical and psychological research and literature that is necessarily limited by particular disciplinary concerns, focused around, for example, bodily symptoms, or learned behavioural responses. I argue that a critical geographical perspective is better placed to take account of the importance of wider social contexts and relations, and can give a fully spatialized account of the disorder more faithful to the way sufferers actually describe their experiences. It may thus go some way toward providing agoraphobic subjects with the means to re-conceptualize, re-articulate and possibly re-structure their problematic relations with the social world.

Theoretically, I would suggest that an investigation of agoraphobia can address and expand some of the core themes of human geography, such as geographies of gender, of women's fear, of disability and impairment, mental health and illness, and of embodiment. By drawing attention to some of the more unusual ways that people relate to each other, and to their environments, we can illuminate some ordinarily taken for granted aspects of personal geographies. People and places are intimately interconnected, they shape and affect each other, and I argue that a condition like agoraphobia emphasizes, exaggerates, and has the potential to elucidate this dialectic.

The voices of agoraphobic sufferers that emerged through my ethnographic interventions have impacted substantially on my textual approach to the subject and subjects of agoraphobia. This reflexive process had effects that ramified far beyond methodological considerations (see chapter 2) affecting both the theories used and even the final structure of the book itself. So, although the general trajectory of this book is *from* theory *to* therapy it also records some of the revisions and re-envisioning of these theoretical insights in the light of my increasing familiarity with subjects' own stories and through the need to relate such abstractions to their everyday experiences.

In each chapter, theoretical problematics and experiential stories are identified and brought together to explore certain aspects of agoraphobic experiences. Some chapters deal with 'philosophical' issues that seem very far indeed from most people's everyday concerns, yet in every case I try to link these theoretical considerations with sufferers' embodied experience of a world perceived by them to be an intensely problematic place. While each successive chapter calls upon its predecessor to elucidate the nature of agoraphobia, each also has a quite different thematic dimension. Each explores an issue that emerged through respondents' comments as related or relevant to their condition in ways that may not be immediately obvious (for example, stories about supermarkets, or about pregnancy). In this way, chapter by chapter, the book also follows what might be termed the 'adventures of the dialectic' between people and places as previously unforeseen aspects of this relationship are incorporated into an increasingly grounded theoretical framework.

However, before this process could even begin I faced an initial difficulty in choosing a theoretical starting point capable of being adopted and adapted to account for the relation between agoraphobic subjectivities and social spaces. This was made doubly difficult by the fact that, as respondents' stories made obvious, agoraphobia is not a definitely bounded condition but an amorphous, complex and varying 'cluster' of fears experienced in different ways and in different social circumstances by sufferers (Marks 1987: 324). Nonetheless, despite the intensely personalized nature of each experience, and the apparent lack of a single causative factor, research reveals certain networks of similarities running through this condition. For these reasons Isaac Marks, a clinical psychiatrist writing prolifically for professional and lay audiences on this subject, is amongst those who prefer to refer to agoraphobia as a *syndrome* (1987, esp. ch. 10; see also Gournay 1989).

One way of envisaging the sheer diversity of experience and 'stories' encountered in my studies on this agoraphobic syndrome is in terms of Wittgenstein's notion of a 'family resemblance'. This conception enables us to avoid the tendency to reify particular aspects of the agoraphobic experience as 'essential' characteristics present in all cases. Wittgenstein points out that there does not have to be a single common identifying feature in order for us to recognize the relationship between family members. We recognize their affinity by the various networks of resemblances between them in terms of items as varied as 'build, features, colour of eyes, gait, temperament, etc. etc.'. So too, when comparing the experiential accounts of agoraphobic subjects, we find 'a complicated network of similarities overlapping and criss-crossing: sometimes overall similarities, sometimes similarities of detail' (Wittgenstein 1988: 66). Each has their own nuances, their own phenomenology (vertiginous, visual, bodily, emotional), aetiology (architectural, existential, social, psychological) and so on. We may find, for example, agoraphobic *tendencies* to respond to certain kinds of spaces and places in one (perhaps fearful and avoidant) way rather than in another (more 'positive' or protected). We will not find, however, universal reactions to any location or situation, not even those as apparently clear-cut as 'homes' or 'funfairs'.

Crucially, this complexity and diversity of experience strongly suggests that the study of agoraphobia would benefit from the kind of multi-faceted *theoretical* (geographical, sociological, philosophical) as well as methodological approach employed here. In this way we might tease apart some of the constitutive fibres of the agoraphobic syndrome before we attempt to reweave them into a meaningful pattern. This is precisely what the book attempts to do chapter by chapter. Just as 'the strength of the thread does not reside in the fact that some fibre runs through its whole length, but in the overlapping of many fibres', so the explanatory strength of any eventual theoretical synthesis will lie in extending the 'concept of [in this case, agoraphobia] as in spinning a thread we twist fibre on fibre' (Wittgenstein 1988: 67).

Perhaps the most persistent pattern that emerged as one story became woven with another was a picture of agoraphobia as a disorder that problematized sufferers' experience of themselves as bounded and well-defined entities, as individual(ized) subjects. This experience is crystallized in the panic attack(s) that all respondents (without exception) had experienced in some form prior to the onset of their condition (see chapters 1 and 2). Panic, as I will argue throughout this book, is experienced as an unbearable attack on one's sense of self in space, constituting an unmitigated existential threat. That is to say, it destroys one's sense of relatedness to other people, and locatedness in place, alienating the subject from the practices of everyday life.

In describing these attacks and their subsequent repercussions sufferers frequently utilized complex spatial imageries that referred to their felt inability to control the boundaries between 'inner' subjective and 'outer' social worlds. Many spoke of the way space seemed to intrude on their consciousness and/or a feeling of their own self-identity's dissolution and dispersion, of a confusion, vertigo and intense anxiety that emanates from the sudden feeling of lacking a secure ground for their existence in the face of panic. Concentrating on the phenomenology of what appear to be accounts of a 'boundary crisis' seemed to offer a novel geographic approach to agoraphobia as a disturbance of subjects' otherwise taken-for-granted ability to mediate the relation between subject and social space. It also suggested that a possible theoretical starting point for thinking through this relation might be found in the accounts of existential anxiety presented by thinkers like Kierkegaard, Heidegger, Sartre, and Merleau-Ponty.

Existentialism, as a philosophical tradition, not only deals directly with the relation between anxiety (*Angest* for Kierkegaard, *Angst* for Heidegger, *Angoisse* for Sartre) and existence, it also offers detailed accounts of the inseparable relation between person-hood and place, what Heidegger articulates in a single term as our 'Being-in-the-World' (Heidegger 1988).[1] For Heidegger human existence is characterized as *Dasein* – quite literally as a '*there*-being', where 'being' is a verb – an activity, namely, existing. Thus the relation between place and person is not characterized in terms of those between a ready-formed subject and their 'objective' environment. Rather, the nature of our existence is such that we are always, already, gathered together in (as) a locus. 'Being-there' is more like 'being-*the*-there'. In Giorgio Agamben's (1991: 5) words, 'being its own *Da* (its own there) is what characterizes *Dasein* (Being-there)'. Or, we might say, human existence is about being able to constitute oneself *as* (in) a specific place. This obviously opens up all sorts of possible connections for geographies of the subject. And, while Heidegger's own work does not feature prominently in this book, the processual nature of this work also traces and testifies to my own intellectual engagement with the existential tradition, from Kierkegaard through to those heavily influenced by Heidegger, like Sartre and Merleau-Ponty.

To reiterate then, while each substantive chapter has a separable thematic content, the book generally moves from theory to therapy. It also follows the unfolding development of the existential and phenomenological tradition in its movement from the more esoteric and subjectivist existentialism of Kierkegaard, through the socially grounded work of Sartre, to the explicitly spatial and embodied and inter-subjective account of existentialism presented by Merleau-Ponty. In this way the book also moves from abstract philosophical arguments about the human condition *per se* to a more nuanced feminist geography capable of accounting for the diversity of experiences of agoraphobia and its gendered relations to particular times and places.

Chapter 1 of this book is devoted to a critical review of the various bodies of literature deemed relevant to this enquiry, a review intended to highlight a number of gaps that this project seeks to address. By critiquing existing understandings of agoraphobia, I will begin to introduce the kind of theoretical and conceptual resources that are required for this project, feminist, phenomenological, existentialist, etc., a discussion that prepares the ground for presentation of my experience of actually *doing* research on agoraphobia in chapter 2.

Chapter 2 focuses in particular on the negotiation of one's positioning in relation to respondents. My experience reveals that, no matter how carefully such issues have been thought out in advance, the subjects of our research may have ideas and plans that do not coincide with, and effectively thwart, our own. The chapter explains how one group in particular presented a degree of 'resistance' to my role as temporary group member, resistance that was enacted through humour, but which was taken seriously enough to prompt the emergence of an alternative, 'processual' approach to the project. My manner of involvement with groups was thereafter a continual 'work in process', my 'method' an evolving response to subjects' 'jocular' policing of the boundary between insider/outsider status. The course of action this chapter outlines was thus intended to enact (and accept) positional change according to shifting dynamics between and among researcher and researched.

As outlined above, the shape of this book partially embodies the processual unfolding of my theoretical interactions with the subject(s) of agoraphobia. Chapter 3 presents my engagement with one predominant experiential theme in relation to an early existentialist appraisal of human being. In particular it focuses on the phenomenology of the panic attack (characterized in terms of an existential crisis) and its relation to social space. The social spaces that recurred most frequently in women's narratives were, perhaps unsurprisingly, those associated with the everyday (and gendered) activity of shopping. This chapter therefore draws on and develops existentialist insights to help articulate something of the inherently spatialized nature of agoraphobia in order to investigate why consumer spaces, and shopping malls in particular, should present such a substantial threat to many sufferers' boundaries and self-identity. Despite what one would imagine to be the relatively trivial place of 'shopping' in an existentialist worldview, I argue that the (psychological[2]) works of Kierkegaard can be utilized to explain why consumption should present sufferers with difficulties. Sufferers' debilitating experiences of such consumerist 'pathoscapes' are contrasted with their largely positive experience of the home, a space often perceived to be protective, rather than corrosive, of the agoraphobic's sense of self. This contrast is intended to underline one inescapable outcome of this research – that selves and spaces impact on and interact with each other.

Chapter 4 continues this existentialist exploration of the self/space dialectic, and shifts the theoretical focus from Kierkegaard to Sartre, examining the latter's account of the centrality of Angst to our being-in-the-world, again in close relation to respondents' own tales of social and spatial anxieties. In particular, it considers subjects' fears of being unable to maintain self-identity in the face of the 'look' of others, and their means of handling such experiences. In a move toward recognizing the importance of therapy for this book, it is argued that the techniques of 'front management' and 'self-presentation' posited by Goffman can be regarded as a form of coping mechanism for dealing with the anxieties of social existence outlined by Sartre. Further, that one, potentially useful way of understanding agoraphobia might be as a failure to utilize these coping mechanisms. By thus bringing sociological insights to bear on this existentialist approach, the chapter suggests that sufferers might benefit from re-learning how to use particular coping mechanisms to strengthen their weakened and fragile boundaries. In this way, it underlines the increasingly socially grounded nature of existentialist approaches to being human in contemporary society.

In chapter 5, I call on Merleau-Ponty to further extend the exploration of the nature of agoraphobic fears of social spaces, focusing particularly on the phenomenological implications of these fears for spaces of the self. Merleau-Ponty highlights both the mediating role of sensations in acquiring a sense of identity, and the importance of recognizing the social (as opposed to merely individual) construction of lived space. Perhaps even more importantly his work critiques and provides an alternative to dominant Cartesian presuppositions of mind/body dualism, for Merleau-Ponty the self is not only social but is inextricably embodied.

This chapter thus provides a vital link between existential accounts of being-in-the-world and 'being embodied' that become the focus of theoretical and therapeutic concerns in this and later chapters.

The second part of chapter 5 addresses the potential relevance of this approach for sufferers from agoraphobia through a specific case study that serves to emphasize the embodied nature of agoraphobic experience. Inevitably, this strategy means that we come to know 'Linda' rather better than the other respondents in this book. However, the intention behind this approach is not to claim that Linda is somehow a typical agoraphobic (since no such 'typical' person exists) or to marginalize other respondents' accounts – some of whose experiences are also presented in some detail throughout the book. Rather, Linda's individual story in this chapter is intended to provide a particular insight into her agoraphobic experience(s), helping to draw out and illustrate the implications and importance of a 'spatial understanding of mental health', the links between phenomenological and existentialist approaches, and the limitations of Cartesian mind/body dualism. Focusing on the personal narrative of just one individual also allows therapeutic concerns to begin to emerge from theory through what might be termed an 'existential analysis'.

Chapter 6 further develops the book's increasingly geographical, and explicitly feminist take on existential phenomenology. Calling on Sartre's insights regarding objectification (chapter 4) and the account of Merleau-Ponty advanced in chapter 5 it seeks to draw out some of the project's thematic fibres relating to a gendered phenomenology of embodiment and bodily boundaries. In particular it explores possible connections between the gendered 'conditions' of pregnancy and agoraphobia by linking my own work with Robyn Longhurst's recent research on the embodied experiences of pregnant women in public places. There are, I suggest, commonalties and possible synergies between pregnancy and agoraphobia since both can be interpreted as entailing severe disruptions to bodily boundaries and might usefully be seen as experiential intensifications of the problems associated with maintaining a stable sense of self-identity by women. In both cases, subjects experience a heightened awareness of socially problematic aspects of womanhood, and raised sensitivity to the feeling of being 'out of place' in the public sphere.

The last chapter completes the grounding movement from theory to therapy and draws us back into the research context itself, namely the self-help groups that remain the methodological focus of the project. The purpose here is to begin to suggest ways in which a feminist geography of agoraphobic existence might elucidate generally overlooked aspects of issues regarding sufferers' treatment. In particular, it suggests how an analysis of self-help resources from the perspective developed within this book might offer new insights and different possibilities for recovery. While self-help resources are often overlooked and undervalued in academic studies they actually constitute an important part of sufferers' attempts to mitigate their condition. However, such texts are far from unproblematic, since they often presuppose a notion of the 'self' that is overly, and often overtly, reliant on the individualistic, yet internally dualistic Cartesian concepts criticized by this book. By critiquing this limited and limiting Cartesian notion of self, and questioning what self-help resources of a more inclusive hue might look like, I offer an account of subjectivity more sympathetic to the 'unusual' spatial relations encountered here. This perspective is also less likely to pathologize agoraphobic experience. This chapter thus begins to rethink spaces in, of and around the self that have been explored and entwined throughout the book, and highlights the unacknowledged centrality and influence of existential philosophies for 'real world' situations.

In 'conclusion', this geographic story attempts to draw together the various strands that have contributed to this re-telling of tales about agoraphobic life-worlds. There is no singular 'meta-narrative' here, no intention to encompass the totality of individuals' experience within a restrictive and reductive framework, but rather, a collage and collection of 'minor' (Katz 1996) stories told around particular themes from particular perspectives, one that amounts to a (re)composition of agoraphobia, and potentially also, of agoraphobic subjectivities. There may be no clear end in sight to many of the tales told here. The theses I put forward are only steps on the way toward an understanding of agoraphobia, steps that will almost inevitably include the occasional wrong turn. As for agoraphobics themselves, their lives frequently seem to be part of an ongoing saga, one that I have come to realize, often requires great heroism in fearful circumstances.

But what is important in any tale is that it should offer some hope, a hope that can only emerge by providing insights into the nature of our being-in-the-world, into what Hannah Arendt (1958) calls The Human Condition. Agoraphobic narratives express both something general about this condition, in terms of exposing our underlying existential anxieties and something very specific about the gendered geographic particularity of women's positioning, their 'there-being' in the modern world. Agoraphobia makes explicit Arendt's (1958: 184) claim that 'nobody is the author and producer of his [sic] own life story'; it highlights the implicate order of our embodied existence in social spaces. But in doing so it also offers the hope that through understanding agoraphobia's implications the possibility arises of an analysis of, and a therapy for, the self (and society) that might help alleviate some of the symptoms of this particularly debilitating condition.

Notes

[1] 'That which anxiety is anxious about is Being-in-the-World itself' (Heidegger 1988: 232).
[2] '*The Concept of Anxiety* and *The Sickness unto Death* are aptly enough called Kierkegaard's psychological works' (Hannay 1982: 157).

Chapter 1

What in the World is Agoraphobia?

Introduction

The condition of agoraphobia is said by one of the foremost clinical researchers on the subject, Isaac Marks (1987: 323), to be 'the most common and most distressing phobic disorder seen in adult patients'. Since the late 1970s, there has been a 'surge of interest' (Chambless 1982: 1) in agoraphobia and panic among clinicians, and publications on the subject often begin by drawing attention to its contemporary prevalence (McNally 1994: vii; Hallam 1992: 114). Rachman (1998: 115) has gone so far as to describe agoraphobia as 'the prototypical modern neurosis', and the centrality of related anxieties to contemporary life has not gone unnoticed beyond clinical circles. Feminist geographer Linda McDowell, for example, points out (1996: 30) that 'anxiety is a central theme in a great deal of current work', while historians Sarah Dunant and Roy Porter (1996) explore contemporary mental life in what they see as an *Age of Anxiety*. Social theorist Jackie Orr (1990) describes panic disorders as arising in response to 'postmodern existence', and philosophers Kroker, Kroker and Cook (1989) speak of 'panic postmodernism'.

In this chapter I intend to review the various bodies of literature of relevance to agoraphobia and panic from a particular, and necessarily partial, perspective, presenting a discussion structured around certain themes. Some of these, relating, for example, to gender, embodiment and boundary issues, have been a constant preoccupation throughout this project, even influencing my initial interest in agoraphobia as a subject for research. Others have emerged as being of crucial importance as a result of the research process itself, via a combination of and interaction between my theoretical and ethnographic engagement with the subject(s) of agoraphobia. By reviewing the literature in relation to these influential themes I intend to re-present the genuinely interdisciplinary 'frame' or context within which this book was shaped. It is hoped that this discussion helps the reader to position themselves relative to the project's perspective, and clarifies its place in and contribution to contemporary geographical research.

First, I will outline **definitions** of agoraphobia, drawing particularly on clinical literature that accounts for, for example, its precipitants and prevalence, and attempt to expand overly narrow psychological definitions through references to the social and spatial aspects of agoraphobia. This task extends to the following section, in which I consider the **phenomenology** and **embodiment** of the condition, examining the various ways in which it is actually experienced by sufferers, and how this experience is often marginalized in clinical accounts. I make some tentative suggestions about how feminist theories and methods might help rectify this failing; suggestions that are taken up in detail later in the book. I also indicate how an examination of the phenomenology and embodiment of agoraphobia has the potential to expand and augment feminist insights into geographies of 'the body'.

Since agoraphobia is, as the definitions indicate, a spatially mediated anxiety, the problematic nature of **social space** is obviously crucial to an understanding of agoraphobic experience. Sufferers' inability to engage with the social arena obviously imposes severe restrictions on their life-worlds, and I next consider how and why certain kinds of spaces and places are considered so difficult and threatening, drawing on geographical, sociological and architectural writings. Additionally, geographical research on health, illness, and disability has produced valuable understandings of the potentially disabling nature of social space, and I indicate how this project can benefit from such findings, as well as contribute to academic and social awareness of the importance of the topic.

It is a significant and pervasive concern of this book that the overwhelming majority of agoraphobics are women, and the final section of this chapter examines perspectives on the **gendered incidence** of the disorder. Existing accounts of women and agoraphobia, both clinical and non-clinical, fail to give satisfactory explanations as to why women are more susceptible to the disorder than men. This section suggests that feminist geographical perspectives, influenced by phenomenological theory and, to an extent, object relations psychoanalysis, can go some way toward helping us conceptualize and comprehend the gendering of agoraphobia. Here agoraphobia is characterized as a disorder that affects sufferer's sense of **bounded** identity. The review thus aims to emphasize the nature and importance of links that this project makes between these various bodies of literature and how the book as a whole, in turn, feeds back into separate strands of geographical research.

Definitions

- agora – Gk. Hist. An assembly; a place of assembly, esp. a market place.
- phobia – (A) fear, (a) horror, (an) aversion; esp. an abnormal and irrational fear or dread aroused by a particular object or circumstance.
- agoraphobia – [f. Gk. AGORA n. + -PHOBIA.] Irrational fear of open spaces.

(Shorter Oxford English Dictionary)

Contrary to popular opinion, the condition of agoraphobia actually relates to public and social, as opposed to 'open' spaces. Although such widely held misconceptions are supported by dictionary definitions of the disorder, the *Shorter Oxford English Dictionary's* notes on the term's *origins* do in fact provide substantial clues as to the realities of agoraphobia for its many sufferers. Exactly how many sufferers there are, however, is a matter of some debate, and estimates range from between 6 per 1000 (Mathews et al. 1981: 12), to 6 per 100 (NIMH in McNally 1994: 26).

In Lewis Mumford's account of *The City in History*, the agora is characterized as an essential part of ancient Greek city life. It is 'the formal marketplace ... [but also] the convenient open space where the elders met, big enough for the whole village to gather in' (1991: 158). It was 'laid out deliberately to serve alike as market, as place of assembly, and as festival place; and though one part of the agora was often reserved for housewives, *the agora was pre-eminently a man's precinct*' (1991: 177, emphasis added). Significantly, to this day, it is women who tend to have difficulties of a phobic nature with such public and arguably still masculine realms.

The first usage of the term agoraphobia can be dated to 1871, when it was introduced by German psychiatrist Karl Otto Westphal to describe the fearful condition he found common to three of his patients. Agoraphobia was the term he chose to capture 'the impossibility of walking through certain streets or squares, or the possibility of doing so only with resultant dread of anxiety' (quoted in Marks 1987: 323. See also Marks 1975; 1987a). In the previous year, the term *platzschwindel* – dizziness in public places – was coined by Benedikt to describe what is surely another incidence of agoraphobia (Mathews et al. 1981: 1). Psychiatrist Diane Chambless (1982: 1) points out that descriptions of what she claims to be the same disorder were provided in the seventeenth century by Burton in his famous *Anatomy of Melancholy* (although she provides no textual evidence or specific references to justify this claim). Since then, she claims, agoraphobia 'has suffered from a plethora of labels: anxiety hysteria, phobic anxiety depersonalization syndrome, *peur d'espace, horreur de vide*, and most recently, endogenous anxiety'. To this extensive list of labels Isaac Marks (1987: 325) adds *platzangst* and kenophobia, before concluding, 'agoraphobia is the least unsatisfactory of those [labels] proposed so far'. Despite Marks' recognition of the numerous descriptive terms applied to 'the same mental event' over the years, he states that in fact, 'the phenomenology of agoraphobia has not changed since it was described by Westphal in 1871' (1995: 107). While the experience of sufferers themselves *may* have remained the same, health care professionals' descriptions and diagnostic criteria continue to evolve.

In the contemporary context, the definition of agoraphobia employed in clinical practice and publication tends to be drawn from the American Psychiatric Association's *Diagnostic and Statistical Manual of Mental Disorders*. The most recent edition (DSM IV) was published in 1994, and its concept of agoraphobia differs in several important ways from the past two editions. DSM III was the first edition to distinguish and provide separate criteria for agoraphobia, as distinct from simple phobia and social phobia (defined below). Prior to this 1980 publication, all three conditions had been included under the general category of phobic neurosis. DSM III introduced a new distinction between agoraphobia with or without panic attacks, and between generalized anxiety disorder (GAD) and panic disorder (PD). Previously, both had been classified as anxiety neurosis, and the division was newly drawn to highlight the presence or absence of panic (McNally 1994: 4).

Further differences accompanied the publication of the next revised edition (DSM IIIR) in 1987. McNally (1994: 4/5) states that '[b]y then it had become apparent that most cases of clinical agoraphobia developed in the wake of panic, and rarely otherwise', and that '[t]hese observations suggested that agoraphobia does not constitute a distinct syndrome in its own right, but instead emerges as a consequence of the core disorder of panic'. The former condition was, therefore, subsequently considered a sub-set of the latter, and clinicians currently acknowledge panic disorder *with* agoraphobia (PDA) or panic disorder *without* agoraphobia (PD) (Baker 1995; Gelder et al. 1996).

Consequently, the name currently applied to the condition examined in this book is, as defined by (DSM IV) criteria, Panic Disorder with Agoraphobia. The disorder involves 'recurrent unexpected Panic Attacks' (APA: 402/3), followed by

> persistent concern about having additional attacks [and/or] ... worry about the implications of the attack or its consequences (e.g., losing control, having a heart attack, 'going crazy') [and/or] ... a significant change in behaviour related to the attacks.

Crucially, and predictably, the condition must also involve the presence of agoraphobia, which is defined (APA: 396/7) as;

> anxiety about being in places or situations from which escape might be difficult (or embarrassing) or in which help might not be available in the event of having an unexpected or situationally predisposed Panic Attack or panic-like symptoms. Agoraphobic fears typically involve characteristic clusters of situations that include being outside the home alone; being in a crowd or standing in a line; being on a bridge; and travelling on a bus, train, or automobile.

DSM IV further stipulates that these situations are then avoided 'or else are endured with marked distress or with anxiety about having a Panic Attack or require the presence of a companion', and that the anxiety or phobic avoidance is not better accounted for by another disorder, such as

> Social Phobia (e.g., avoidance limited to social situations because of fear of embarrassment), Specific Phobia (e.g., avoidance limited to a single situation like elevators), Obsessive-Compulsive Disorder (e.g., avoidance of dirt in someone with an obsession about contamination), Posttraumatic Stress Disorder (e.g., avoidance of stimuli associated with a severe stressor), or Separation Anxiety Disorder (e.g., avoidance of leaving home or relatives) (APA: 402/3).

Despite the specificity of the definition outlined, this book will henceforth refer to its central subject simply as 'agoraphobia'. While this strategy is adopted primarily for purposes of simplification, it is justified by the fact that only one form of the disorder has been encountered, persistently and exclusively, throughout this research. The agoraphobic condition confronted is one that, despite its aetiological and phenomenological differences, *always* involves experience and fear of panic (but see Bankey (2001) on alternative usage of the term).

As mentioned above, the vast majority of agoraphobia sufferers – around 89 per cent – are women (Clum and Knowles 1991). This issue will be dealt with at length in the following section of this chapter, but does in fact constitute a core concern of the book as a whole. For those (women and men) who develop agoraphobia, the onset tends to occur between the ages of eighteen and thirty-five. Burns and Thorpe (1977) found the mean age of onset to be 28 years, Buglass et al. (1977) 31 years, Ost (1987) 28 years, and Solymon et al. (1986) 25 years[1], and the condition tends to be chronic. Michelson and Ascher (1987) found the average duration of agoraphobia is eleven years. There is, however, evidence that a degree of agoraphobia may remain with the sufferer for life. Kevin Gournay (1989: 1), for example, reports that not one of the 132 patients who took part in his long term trials achieved 'anything amounting to a total abolition of symptoms', and according to Chambless and Goldstein (1980: 120), '[t]ransient improvements aside, once the symptoms begin, they generally persist over a life-time with fluctuating severity'. Many sufferers will themselves attest to the fact that one can never be entirely 'cured' from agoraphobia, but must rather learn to cope with it on a continuing, daily basis (see also Chambless and Goldstein 1982). To this end, sufferers frequently adopt certain tactics and strategies, which, as I argue in chapter 4, function to reinforce weakened boundaries and to discourage social contact. Commonly used methods of partially managing anxiety, such as wearing dark glasses or holding a newspaper, have been noted by clinicians (e.g. Marks 1987: 337-9).

Clinical research on the issue of particular causes has not been entirely conclusive, but there is evidence that higher than usual levels of stress tend to be experienced prior to the onset of agoraphobia (Barlow 1988: 32; Michelson and Ascher 1987: 217). While the nature of such stressors and negative life events vary between individuals, Rachman (1998: 118) suggests, in a provocative insight to be taken further by this book, that what they may have in common is an ability to 'undermine one's sense of safety'.

With regard to issues of treatment of agoraphobia, reviews of recent literature suggest that the form most likely to be successful is a multi-faceted approach capable of recognizing and tackling each problematic aspect of the disorder (Rachman 1998). Cognitive Behavioural Therapy (CBT) advocates Barlow and Cerny (1988: 28) criticize earlier treatment models they describe as 'one dimensional' (including narrow psychological and biological approaches), but recognize that drug treatments may at times still be appropriate as part of a larger programme of treatment. Referring to the currently dominant (DSM IV) view of agoraphobia as an outcome of panic disorder, they argue that management of panic itself should be treated as a matter of priority. This view is based on the understanding that agoraphobic subjects are *primarily* afraid of panic – of fear itself, and not particular places. Thus, it is thought that if the subject can learn to control their experience and fears of panic, issues of agoraphobic avoidance will be resolved much more easily, by the traditional means of controlled excursions into difficult places (Rapee 1987). Such treatment models do, however, tend to assume a simplistic, liberal humanist version of the self, one that is predicated on the subjugation of one (female) side of hierarchic dichotomies. While this claim is elaborated, and the implications of such models of treatment discussed at length in chapter 7, 'Enlightenment' conceptions of subjectivity are problematized throughout this book.

It seems then that while the definitions and descriptions of agoraphobia to be found in the clinical literature acknowledge much of the disorder's complexity (in terms, for example, of its interconnections with other phobias), its representation tends to be simplistic in certain important respects. In particular, the model of the subject that such writings employ tends to be somewhat philosophically naïve. Also, while recognizing that agoraphobia is socially and spatially mediated, understandings of what this might *mean* for sufferers and how social and spatial theory might be of benefit are, as one might expect, limited. In the chapters that follow, I want to suggest that some of these limitations might be at least partially rectified by the *care*ful inclusion of sufferers' own voices (conspicuously absent from clinical accounts) and resolved further with reference to a sympathetic theoretical (feminist, geographical and phenomenological) perspective. The book thus aims to refine and improve existing 'definitions' of agoraphobia, by offering a more spatialized and experientially influenced perspective that can better account for and understand the nature and (potentially pathological) implications of person/place relations.

In the following section, I want to begin the task of constructing a more 'intimate' picture of experience and meanings of agoraphobia, by considering its implications for spaces of the self. I will first provide a brief, clinical account of the 'defining' experience of agoraphobia – panic – before broadening the discussion to include non-clinical perspectives. While the definitive clarity of the former is not mirrored in non-clinical writings on agoraphobia, many of the latter are immensely valuable precisely because they eschew clinical precision in favour of socio-spatial and experiential significance, and cultural/historical context (de Swaan 1981).

Phenomenology and the Embodiment of Agoraphobia

The phenomenological aspects of agoraphobia of overwhelming and terrifying significance for its sufferers relate to the experience of panic. A panic attack is defined by DSM IV (APA: 395) as being:

A discrete period of intense fear or discomfort, in which four or more of the following symptoms developed abruptly and reached a peak within ten minutes:

1. palpitations, pounding heart, or accelerated heart rate
2. sweating
3. trembling or shaking
4. sensations of shortness of breath or smothering
5. feeling of choking
6. chest pain or discomfort
7. nausea or abdominal distress
8. feeling dizzy, unsteady, lightheaded, or faint

9. derealization (feelings of unreality) or depersonalization (being detached from oneself)
10. fear of losing control or going crazy
11. fear of dying
12. paresthesias (numbness or tingling sensations)
13. chills or hot flushes.

Throughout this book, my interest in symptoms such as these relates to the effect on sufferers' experience of embodiment and environment, that is, the ways in which symptoms of panic problematize subjects' understanding of themselves as clearly and distinctly separate from the world around them. I contend that the symptoms listed above cause sufferers to question their ordinarily taken for granted sense of themselves and their bodily boundaries, and that exploration of such phenomena is of crucial importance for understanding experience and meanings of agoraphobia. To warrant such interpretations, and expand existing views, it is clearly necessary to move beyond the clinical literature. In the discussion that follows, it will become apparent that sufferers' own accounts can powerfully, if alarmingly, bring clinicians' dispassionate descriptions to life.

Derealization and depersonalization (listed above, no. 9), for example, are commonly understood to involve strange and disturbing feelings relating to both 'inner' and 'outer' reality. During such experiences, according to Marks (1987: 342) 'one feels temporarily strange, unreal, disembodied, cut off or far away from immediate surroundings'. (See also Chambless & Goldstein 1982.) We can usefully compare this portrayal with the following quotation (taken from a self-help book) in which one sufferer describes personal experience of the phenomenon. The tenor of their description provides deeper insight, and puts a disturbing twist on our understanding of what it means to feel 'faint':

> [S]uddenly everything around me seemed unfamiliar, as it would in a dream. I felt panic rising inside me. [Pause] I felt totally unreal. [Pause] Two people came along and I almost stopped them to ask if they could see me – was I really there (Melville 1979: 13).

In another self-help book, Law (1975: 53) communicates further the intensity of the experience:

> Suddenly, sensations stronger and stranger than any I had previously known charged through my body. [Pause] I turned and twisted and tried to remove the devilish feelings that possessed me. [Pause] This evil thing threatened to rob me of my identity. [Pause] Everything seemed to disintegrate.

This sufferer feels that reality is 'tearing in on him', as if he were receding into an abyss, with a mighty hand trying to 'squeeze the breath, and every particle of self identity from [him]' (Law 1975: 72). This quotation, and the one above, reveal that panic entails a horrendous sense of dissolution of self into one's environs, and a simultaneous feeling of invasion by one's surrounds. Panic severely disrupts our unconscious sense of ourselves, throwing the relation we have with our bodies, our selves, into profound question. As the (DSM IV) list of symptoms itself reveals, some aspects of agoraphobic panic can be described as of an *existentialist* nature (esp. nos. 10 and 11), causing sufferers to question the very grounds of their existence, to fear outright loss of control over themselves, or the ultimate loss of death itself.

Existentialist philosophies have thought that such anxiety (*Angst, Angoisse*) provides a key to understanding the human condition. In his introduction to Søren Kierkegaard's *The Concept of Anxiety* (1980), Reidar Thomte writes that

> [f]ear is a threat to the periphery of one's existence and can be studied as an effect among other effects. Anxiety is a threat to the foundation and center of one's existence. It is ontological and can be understood only as a threat to Dasein (xvii).

Anxiety then, on this view at least, disrupts the very essence of our 'being-in-the-world'. We might expect then that existential philosophies like Kierkegaard's might be able to shed some light on the nature of agoraphobic anxiety and panic. Interestingly clinical and (especially) experiential depictions of such episodes are in many ways reminiscent of certain salient portrayals of existential *Angst*. *Nausea*, for example, Sartre's (1965) semi-autobiographical novel, describes the central character's increasing 'alienation' from various objects, and culminates with this disturbing account of an encounter with a tree in a park:

> It had lost its harmless appearance as an abstract category: ... the root, the park gates, the bench, the sparse grass on the lawn, all that had vanished; the diversity of things, their individuality, was only an appearance, a veneer. This veneer had melted, leaving soft, monstrous masses, in disorder – naked, with a frightening obscene nakedness (183).

For *Nausea's* protagonist Roquentin (and by extension, Sartre[2]), even parts of his own body could appear bizarre and disturbing, his hand as a 'fat slug' resting on the table before him, his face, some alien thing in the mirror.

Following the onset of anxiety, of 'fear and trembling' (Kierkegaard 1985) as a way of life, there is a loss of *trust* in the integrity of the body. The accounts of agoraphobic respondents (see chapter 2) utilised throughout this book repeatedly emphasize that the individual sufferer fears being betrayed by their body, and is constantly on the look out for breaches of bodily boundaries that might be visible to others. These include for example, minor discrepancies such as blushing, trembling or sweating, in addition to more extreme examples, such as screaming out loud, or the loss of bowel or bladder control, manifestations of their fear on the surface of their bodies which would be thoroughly humiliating. The intensely *embodied* nature of these 'emotional' states necessitates a reconfiguration of the subject's conception of herself. The disorder seems to initiate a sense of *separateness* from one's body, not just momentarily, as in the fleeting depersonalization of the panic attack, but in the more general sense of creating a constant anxiety which hovers over the question of our bodily and mental identity.

Precisely because something seems to have *gone wrong* agoraphobic anxiety forces our bodily and mental states to the forefront of our attention; we are no longer 'at one with' our body, and our relationship with it is no longer unconscious or unproblematic. We might, following Drew Leder (1990), speak of a sense of *dys*embodiment. For Leder, the body is ordinarily 'absent', only in the event of some dysfunction does it become immediately present to us, it *dys*appears. Thomas Csordas, accepting Leder's portrayal, characterizes it thus; 'the vivid but unwanted consciousness of one's body in disease, distress or dysfunction is a kind of dys-appearance, a bodily alienation or absence of a distinct kind' (Csordas 1994: 8). In agoraphobia, our sense of embodied subjectivity is inescapably conspicuous, as indeed it is with many other manifestations of chronic illness and disability (Butler and Parr 1999; Dorn 1998; Moss and Dyck 1999). But, given the ease with which we *normally* 'take for granted' our status as clearly defined individuals, as solid bodies which contain (and also constitute) our personalities, it comes as something of a shock to realize that, in Elizabeth Grosz's terms, we might inhabit 'volatile bodies'. For Grosz (1994), mind and body alike are not static or impervious to our surroundings but labile and at least semi-permeable;

> [t]he limits or borders of the body image are not fixed by nature or confined to the anatomical 'container', the skin. The body image is extremely fluid and dynamic; its borders, edges and contours are 'osmotic' – they have the remarkable power of incorporating and expelling outside and inside in an ongoing exchange (Grosz 1994: 79).

When the ability to subconsciously regulate these processes of incorporation and expulsion slip from our control then our very existence is thrown into doubt. 'I was startled awake one day to realize that *I was almost entirely gone*. What woke me was my body, whose very being in the world suddenly shifted and changed everything' (Bordo 1998: 80). The difference between the sufferer and normalcy is epitomized in Bordo's account of her recovery from agoraphobia.

> Being outside, which when I was agoraphobic had left me feeling substanceless, a medium through which body, breath and world would rush, squeezing my heart and dotting my vision, now gave me definition, body, focused my gaze (Bordo 1998: 83).

From a phenomenological position it is unsurprising that anxiety should manifest itself both 'bodily' and 'mentally' since theorists like Merleau-Ponty assume a model of identity which regards the body *as* the self, the 'self expressed' (Merleau-Ponty 1962). Such phenomenological insights continue to influence feminist theory both within and beyond the philosophy discipline, and have a significant and recurring role to play in this book, a significance that stems precisely from the *phenomenological understanding of embodiment and self-identity as inextricably interconnected*. This insight is crucial, I argue, to interpreting and potentially understanding the meaning and experience of agoraphobia.

In contradistinction to the dominant cognitivism of most post-Cartesian philosophy there is something of a consensus among contemporary feminist theorists, geographers included, that the subjectivity or sense of self of each and every individual is thoroughly, absolutely, embodied. This re-emphasis on the body has drawn on numerous sources but, as Bordo (1993: 17) notes, owes as much if not more to the body-politics of second wave feminism as it does to the theorizing of Foucault. Whatever its origins, all agree that this long overdue acknowledgement of the inter-relatedness of self-identity and corporeality opens new questions. As soon as we recognize that the body and not just the mind is 'a medium of culture' (Bordo 1993: 65) then the question of how women's self-identities and bodies are maintained and disciplined in and through the agency of an (often hostile) cultural environment becomes crucial. The question becomes one of the consequences for self-identity of the relative (im)permeability of that subject's boundaries to influences that are partially constitutive of, and yet potentially corrosive to, that identity. These boundaries can themselves no longer be envisaged in entirely physical or mental terms – as dependent, say, upon a 'strong constitution' or a 'weak will' – but as an inextricably complex combination of both. The question that is constantly asked of each of us is, how 'thick-skinned' can/should we be in the presence of others? This question seems to be absolutely central to an understanding of the nature of agoraphobia and also, to the gender specific incidence of agoraphobia in society.

I argue that a feminist approach that draws on phenomenological, as well as existentialist theory, is particularly suited to the aims of this project for a number of reasons. First, it initially requires that we 'bracket' epistemological questions, and 'go back to the things themselves' – the 'phenomenological reduction' described by Husserl as 'époche'. Phenomenology thus allows us to return to, and concentrate on, the accounts of women's actual lived and embodied experiences, to take seriously their own voices and the stories they tell, rather than requiring us to posit specific external causes. This, as Merleau-Ponty's (1962) account of the 'lived body' (or 'self expressed') shows, in no way necessitates that we ignore the influence of the material and social world. On the contrary, it acts as a reminder that this world is always a world-for-us, interpreted and experienced through our positioning within it. The individual's particular perspective is, therefore, of the utmost phenomenological importance, and this approach thus subverts the Cartesian tendency to divide the world into an internal mental realm and an external physical realm. And, as I argue below, this fits perfectly with women's own accounts of their embodied experience of the world (whether phobic or otherwise), which can never be fully or faithfully conceptualized in terms of one side of this dichotomy. (See also Battersby 1998; Leder 1990; Young 1990. An account of phenomenology is developed more fully in chapter 5.)

However, of the feminist theorists called upon in this book, several might align themselves more closely with post-structuralist than phenomenological thought (e.g. Butler 1992; 1993, Grosz 1994 and Longhurst 2000a), and my own inclination toward phenomenology, viewed as more sensual than textual, thus requires some explanation. As Moran (2000) has argued, phenomenological and post-structuralist perspectives are not so different as they might initially seem. Not only does post-structuralism owe a historical debt to phenomenology – particularly in the shape of Heidegger and Merleau-Ponty's writings – that often remains unacknowledged, but both are also conspicuously engaged in a critique of scientistic naturalism.[3] Thus phenomenological references to embodied experiences should not be interpreted as positing a pre-discursive body, a body uninfluenced by and beyond the reach of language's circumlocutions. They do not deny the role and import of the linguistic mediation of, or contribution to, our experiences.

Phenomenology does though explicitly deny that all being is language, that there is nothing beyond the text. While the post-structuralist insistence on the discursive nature of embodiment has much to offer feminist politics and theory, we must make sure that we do not ourselves fall prey to a 'tyranny of the concept'. The fact that language is the main medium of expressing ourselves to others does not mean that the self is nothing more than a narrative structure, that my life can be understood simply as the word become flesh. In calling upon us to pay attention to the phenomena themselves, phenomenological descriptions try to bring to the attention of others, to bring into language, sensory experiences that have as yet been insufficiently or improperly articulated. This difficulty in communication, in finding the right words to express one's embodied experiences, is something of which agoraphobics frequently complain.

In this sense the phenomenological feminism advocated here can be genuinely critical of phallogocentrism – of the power of the word (logos) to order and apportion meaning to women's experiences in a patriarchal society, and comes close to the difference feminism of Luce Irigaray, who is herself frequently regarded as 'post-structuralist' (Irigaray 1985). Irigaray too considers women's embodied experiences as being denied expression in the dominant symbolic order and as providing a (non-naturalistic) 'ground' from which women might speak of the otherwise inarticulable differences of their experiences (Davidson & Smith 1999). One of the explicit aims of this book is to make room for the previously unheard voices of agoraphobic women, to allow unusual stories about selves and spaces to be articulated, and truly appreciated by (phobic and non-phobic) others, all of whom, at one time or another, will have experienced some discomfort in the face of social space. The book aims to help make such strange socio-spatial experience intelligible, and perhaps also more acceptable. That is to say, it aims to challenge the stigma associated with unusual, even 'delusional' geographies (Parr 1998), by rendering them more comprehensible and comparable to 'normal' everyday life. The phenomenological aspects of this investigation thus expand geographical critiques of the (dis)embodied subject, and enhance our understanding of its inherent spatiality.

There has been substantial focus on geographies of embodiment in recent years, with many, often feminist, geographers pointing out that (the) bodies of theory in fact tend to be white, male, heterosexual, middle class, and otherwise normative, not 'normal' (Rose 1999: 249). By this I mean that, despite the frequency with which we encounter such hegemonic, supposedly unmarked and 'neutral' representations of embodiment, they fail to correspond with the experience most of us have of our own and other('s) bodies. In response, such researchers have attempted to re-place predominant and proscriptive approaches to thinking about 'the' body, and bring the live and fleshy diversity of real individual bodies to the theoretical fore. By drawing attention to those that are, for example, young, pregnant, classed, 'raced', sculpted, tattooed, built, bulimic, sexed, queer, disabled, fit, obese, cyber-, grotesque, psychiatrized, old and dead (examples of most, if not all, of which can be found in collections edited by Butler and Parr 1999; Duncan 1996; Teather 1999, and Nast and Pile 1998), geographers aim to initiate conceptual and material change. That is to say, such research can challenge the way we think and thus the way we treat bodies, and so also selves. I want to suggest that studies of agoraphobic embodiment have potentially significant contributions to make to this expanding field. This book thus argues that agoraphobia's experiential and conceptual excess, its stress on the emotional dynamics of embodiment in place(s) constitutes constructive refusal of the dis-embedded bodies of 'tired' and, all to often, untested theory.

One particularly useful sociological approach to embodiment that draws extensively on phenomenological theory is advanced by Williams and Bendelow in *The Lived Body* (1998). Like these theorists, I have found that attempts to extend and apply phenomenological understandings of embodiment to particular lived bodies in particular social contexts benefit considerably from an infusion of the sociological insights of Erving Goffman. For Williams and Bendelow, Goffman's carefully observed and empirically based 'carnal sociology' gives further practical credence to phenomenology, and his work is clearly of geographical significance in its focus on the social settings in which bodily encounters occur. Goffman's work has been inspirational for this book (see especially chapter 4) because of his attempts to elucidate the nature of social anxieties, including stigma and embarrassment, and the tactics by which they are 'managed'.

In a statement that highlights Goffman's relevance for this project, Williams and Bendelow (1998: 56) write that:

> For Goffman, successful passage through public space, whether it be the street, the supermarket, or a busy shopping mall, is both a practical problem and skilful accomplishment for the human agent, involving specific social rules and rituals which facilitate this passage and 'repair' disruptions to the micro-public order of social interaction.

Precisely because of this focus on the largely taken-for-granted 'minutiae' of social encounters in practical(ly) 'everyday' situations – for the non-phobic at least – Goffman's work will be taken up in order to 'flesh out' philosophical treatment of the phenomenal difficulties suffered by phobic selves in social space.

Social Space

In 'Spatialising Feminism', Linda McDowell (1996: 29) emphasizes the need to perform the perhaps obvious but oft-avoided geographical task of *defining* space, and indeed Doreen Massey elsewhere (1993: 142) asserts that a discussion of this topic 'never surfaces because everyone assumes we already know what the term means'. Clearly, however, 'space' is used in geographical literature in a number of diverse and at times conflicting ways. Against this ambiguous background, McDowell begins her attempt at clarification by highlighting the importance of viewing space as 'relational and constitutive of social processes', and crucially, in her view, as inextricably connected with time (1996: 29; see also McDowell 1999, Massey 1992, 1993; Soja 1989). In McDowell's sense 'space' is not Newtonian, it is not an abstract pre-existing void, an empty container waiting to be filled by people, objects etc. Space is something that in Lefebvre's (1994) terms is 'produced' through a (culturally mediated) dialectic between things and people.

Referring to spaces as relational need not imply that they are somehow less concrete, that they lack 'permanence' or 'solidity' and therefore cannot operate to anchor cultural meanings and symbols. This is precisely why McDowell wants to emphasize that space need not be conceived *only* as a 'set of flows' but can also be regarded as a 'set of places' which are specific locations that operate as territories, bounding and fixing these relations in culturally variable ways.[4] Thus, while this characterization allows a sense of space as 'social morphology' (Lefebvre 1994), crucially, it does not preclude discussion of 'sets' of places of potential import for this book – the spaces, for example, of subjects' homes, train stations, or malls – so long as the particular context is explained. For McDowell, 'spatial configurations, connections between places are significant only in the context of a specific question or investigation of specific sets of relationships' (1996: 31). Such relational articulations of space have an important role to play in human geography research and the specific question and particular purpose of this book – to explore and illuminate the nature of certain socio-spatial anxieties – are facilitated by an appreciation of space that emphasizes relations between subjects.

Putting the subject centre-stage is explicit in Gillian Rose's (1999) account of 'Performing Space'. Rose advances an experientially focused understanding of space that takes seriously the possibility of complex and unusual spatial experiences. Moreover, it allows for, indeed 'fits' with agoraphobic narratives of deeply personal and confused, but evidently inter-relational and dynamic experiences of space. Rose argues (1999: 248) that, as with Butler's (1990; 1993) account of gender, 'space is also a doing, that it does not pre-exist its doing, and that its doing is the articulation of relational performances'. Further:

> Space then is not an anterior actant to be filled or spanned or constructed, and to claim it is runs the risk of making a contingent spatial articulation of relationality foundational. Instead, space is practised, a matrix of play, dynamic and iterative, its forms and shapes produced through the citational performance of self-other relations. Which is not to say that space is infinitely plastic. Certain forms of space tend to recur (Rose 1999: 248).

The thinking and making of space is, for Rose, phantas(ma)tically complex. The processes involve emanations from and interactions (including revolts, refusals, inscriptions and investments) between three spatial modalities – the discursive, the fantasized and the embodied – and cannot be separated from the operation(s) of power. In relation to agoraphobic situations, the powerful forms of space that tend to recur are, as we shall see, spaces infused with and reproductive of *fright*ful patriarchal ideologies and conceits. The perspectives, power and (often behind the scenes) performances of others can combine with subjective sensitivities to make space anxious and aversive, domestic, dull and drab, or oppressive and exclusive. And, for the agoraphobic subject caught up in the performance of such antagonistic space, it rarely *feels* like she has much say in its making. Rather, she perceives herself a 'bit player', reacting against and retreating from the marginalizing performances of others.

This conceptualization of space as performative thus goes some way toward setting the context for addressing one question critical for this project. When we ask what causes certain spaces to be *felt*, by certain subjects, in a threatening way, we clearly need to consider the role, activities, intentions – the performances – of others. These performances, combined with one's own (perhaps anxious) contribution, bring particular spaces into being; 'particular performances articulate their own spatialities' (Gregson and Rose 2000: 441). Kathleen Kirby's view of subjective felt spatiality seems at least partially compatible with this account, and also with those of agoraphobics. She writes:

> the problem with space isn't just space;[5] it is the fact that there are other people in it – other people who are creating it, determining it, composing it ... is it surprising then, that space could seem a bit hostile? (1996: 99)

This statement implies that space is charged somehow, populated with the constructions of others. However, most individuals most of the time are quite literally *impervious* to these un-wholesome attributes of space. Space, as Elspeth Probyn points out, 'presses upon bodies differently' (Probyn 1995: 83), to which we must add; bodies, too, press differently on space (Valentine 1996).

Clearly, it is not only feminist geographers who recognize that space is 'filled and animated by the reciprocal relations between individuals' (Vidler 1991: 39). Opening his fascinating excursus on the damaging effect of particular (city, and thus manufactured) spaces on individual health and identity, architectural theorist Anthony Vidler makes the modest assertion that he wants to write a 'small history of modern space'. Following social and spatial theorists such as Benjamin, Simmel and Kracauer, Vidler conceives of space as 'the *expression* of social conditions' (1991: 39), and yet 'reciprocally *interdependent* with society' (1991: 32). In accordance with the feminist geographical approach introduced above, and advanced by this book, Vidler conceives of spaces and selves as mutually constitutive, affecting and at times afflicting each other. We can and indeed do create, and are (at least partially re-) created by, disturbing and pathological landscapes. In some respects then, Vidler's account of self/space relations resembles that of another architectural theorist, who states that agoraphobia, 'like other urban pathologies, tells us something about the way space is constitutive of personality' (da Costa Meyer 1996: 141).

For Vidler, space is 'of course, no more than a cultural and mental construction; for, in historical terms, like the body, or like sexuality, space is not a constant'. Rather (as Victor Burgin also notes) 'space has a history' (Vidler 1993: 31). If, then, we are to produce the particular history he has in mind – a 'sociopsychological history of metropolitan space' – we ought, Vidler claims, to begin in 1871, the year that the 'essentially spatial disease' of agoraphobia was first identified (1991: 34). As for the geographical starting point for this history, we must look to the pathoscapes of modernity's monstrous cities.

> The nineteenth-century city had been understood to harbor dangerous diseases, epidemics, and equally dangerous social movements; it was the breeding ground of the all-levelling masses, of frightening crowds, the insanitary home of millions, an asphalt and stone wilderness, the opposite of nature. The metropolis carried forward all these stigmas, but added those newly identified by the mental and social sciences. It rapidly became the privileged territory of George Beard's neurasthenia, of Charcot's hysteria, of Carl Otto Westphal's and Legrand du Saulle's agoraphobia, of Benjamin Ball's claustrophobia. It sheltered a nervous and feverish population, over-excited and enervated, whose mental life as Georg Simmel had noted in 1903 was relentlessly anti-social, driven by money and haunted by the fear of touching (Vidler 1991: 34).

Georg Simmel was in fact crucial to the project of theorizing the mental life of modernity and the metropolis. On Vidler's reading, he

> staged the 'sensitive and nervous modern person' in front of the backdrop of 'jostling crowdedness and motley disorder', and argued that an inner psychological barrier, a *distance* was essential for protection against despair and unbearable intrusion. The 'pathological deformation of such an inner boundary and reserve', Simmel noted, 'was called agoraphobia: the fear of coming into too close a contact with objects, a consequence of hyperaesthesia, for which every direct and energetic disturbance causes pain'. Simmel's diagnosis was at once spatial and mental: ... [agoraphobia] was a product of the rapid oscillation between two characteristic moods of urban life: the over-close identification with things, and, alternately, too great a distance from them (Vidler 1991: 36/7).

This conception of 'distance', and the implications for subjects of its 'deformation', provides crucial insights into the subject's vulnerability when faced by crowds of others, and links with my own usage of 'boundaries' in this book. The above account makes clear that social spaces of the city at this particular period in time were especially dangerous for 'sensitive individuals', especially given the emergence and increasing importance of 'space conquering techniques', such as radio and railway (Schivelbusch 1980; see also da Costa Meyer (1996) on the connections between agoraphobia and the Industrial Revolution). Such technologies effectively *conquer* spaces *between*, robbing individuals of their comforting and insulating *distance* from others:

For even as electricity overcame distance, it also worked to annihilate that *Denkraum* which was the zone or space of reasoning, destroying the only protection against the experience of phobia, the falling into unreason (Vidler 1993: 47).

The city (and its performance) thus entail a potentially disabling proximity with others, eroding 'any possibility for a stable distance of reflection' (Vidler 1993: 46/7). When robbed of this 'distance', the invasive proximity of others is perceived by sensitive individuals to weaken their boundaries. Significantly, this proximity also creates far greater potential for objectification, the opportunity to be looked *at*, yet not engaged *with* as a subject, as in a conversational exchange. As Simmel explains, social life in the modern city

> shows a great preponderance of occasions to *see* rather than to *hear* people ... Before the appearance of omnibuses, railroads and streetcars in the nineteenth century, men were not is [sic] a situation where for periods of minutes or hours they could or must look at each other without talking to one another (Simmel in Vidler 1991: 41).

This insight, taken up more recently by Goffman, is crucial to understanding the difficulties posed by social space for agoraphobics, who harbour intense fears of others' looks.

Agoraphobia, a condition inseparable from and inconceivable outwith its socio-spatial context, is clearly a source of fascination for Vidler. His account helps us to historically and geographically situate the disorder in relation to the 'social estrangement that seemed to permeate the metropolitan realm', the form of alienation described by Georg Lukacs as the 'transcendental *homelessness*' of the modern world. Agoraphobia was considered as proof that modern cities were in their very form *unhealthy*. Camillo Sitte, in fact, used the disorder to justify his proposals to create 'small, protective city squares' in his native Vienna. Although he was simply mistaken in placing the problem with cities in their wide, open spaces, his recognition of the emotional importance of walls was indeed most perceptive (da Costa Meyer 1996: 145). Around the turn of the century, agoraphobia was considered, along with other newly emerging 'nervous diseases', to be 'characteristic' of city life, 'endemic to urbanism and its effects'. For the modern city's female subjects particularly, its effects were noisome and unnerving; disconcerting effects no doubt related to the fact that modernist space was 'constructed by and for men' (Deutsche 1990).

This book recognizes that the symptoms (and performances) of agoraphobic panic have serious repercussions for individuals' perception (and creation) of the 'external' spaces of their life-worlds as well as the 'internal' spaces of their bodies. Agoraphobic panic can in fact, as I argue throughout, induce a crisis in the 'boundary' between self and space that throws the *existence* of both into doubt. The dissembling, dissipative effects of agoraphobia can thus be seen to create a *garbled* geography, where that 'closest in', the body (Rich 1986) is no longer safely delimited from the outside world. The fearful fantasies projected and performed by agoraphobic panic create a social space corrosive to the subject's boundaries. In an extraordinarily vicious circle of performance / perception of panic, she subsumes the surrounding confusion, and fears that she is 'losing her grip' on reality.

Kirby (1996: 99) writes similarly and insightfully of another spatial phobia, vertigo, that it involves 'a rift between subject and reality, the mobilization of the internal processes of the subject, and a new fluidity of the external realm'. In vertigo 'the internal/external relation breaks down, resulting in a degeneration of interior organization, and finally – one could imagine, in advanced stages – in a confusion of the external order too' (Kirby 1996: 102). There are, then, other affects and effects of garbled spatial experience. Related to and perhaps arising from the problematization of bodily boundaries, is a crisis of *location*. The subject of phobic panic, whatever its 'object', evidently cannot *connect* normatively with her surroundings and feels 'out of place' in a disconcertingly alien world. Phobic accounts have shown that when panic surfaces, it threatens the dissolution of self such that the individual seems, quite literally, 'lost'. She no longer knows *where* she is, both in the sense that her surroundings have become unfamiliar and threatening and in the sense that she cannot *locate or delimit* her own identity, she feels 'spaced-out'. In Bordo's words,

> [e]ach of us [herself and her sisters] has suffered, each in her own way, from a certain heightened consciousness of space and place and our body's relation to them: spells of anxiety that could involve the feeling of losing one's place in space and time (Bordo 1998: 74).

Somewhat similarly, describing his own spatialized 'boundary crisis', Roger Caillois recounts a feeling of 'depersonalization by *assimilation* to space' (Caillois 1984, in Grosz 1994: 47, emphasis added).

Given this crisis state of spatial affairs – of boundaries, distance and location – it is perhaps unsurprising that the subject should eventually become more or less temporarily house*bound*. Her boundless vulnerability means she may have no choice but to remain exclusively within the home, assuming its protective boundaries as reinforcement and extension of the psychocorporeal boundaries of the self. The agoraphobic thus *incorporates* her 'own four walls' as an essential element of her 'ontological security'. Clearly, the association of this sense of security with the home is not a universal experience, but for some at least, the home is 'a truly safe place, the container and springboard for integrated living … the foundation of an ontologically secure existence' (Marilyn K. Silverman, in Bordo et al. 1998: 90; compare Valentine 1998).

Agoraphobic accounts suggest a powerful contrast between meanings attaching to home spaces and the world beyond, a contrast of potential import for this project. Feminist geographers have explored such issues in relation to distinctions between 'private' and 'public' spaces, arguing that the contours of this binary formation are as much a *product* as a reflection of binary constructions of gender (Bondi and Domosh 1998; Dowling 1998; Duncan 1996). Perhaps especially for the agoraphobic, whose condition 'straddles the fault line between public and private' (da Costa Meyer 1996: 141), the boundaries between these spheres are very much fluid and unfixed. This is evidenced by the fact that the home space traditionally conceived as 'private' can itself become threatening and dangerous when, for example, relatives converge at Christmas. Likewise, the 'public' space of the town centre can be experienced as comfortable and safe when no one else is around. (Ironically, given the common and legitimate concerns of non-phobic women for their safety, this is often when darkness provides calm and *cover* from the 'socializing' presence of others.)

The agoraphobic's experience of home can be ambiguous in other ways; for some, home becomes so 'secure' that they are rendered incapable of leaving, and it can thus simultaneously be experienced as both prison *and* asylum. But, despite the *ambivalence* of feelings, it is clearly the space of the home that is overwhelmingly perceived to be safe, the (anti)social space beyond its confines that menaces and disturbs. It thus seems relevant that, despite the achievements of feminisms in expanding women's horizons, the gendered identities of women in general are *still* more likely to be enmeshed with the fabric of the home than those of men. In this context, agoraphobic women's tendency to experience home as a place of safety, of refuge, might be viewed as a retreat into a normatively feminine space where gendered boundaries are protected and potentially reinforced. However, it must again be emphasized that there is no 'universal' experience of agoraphobia, and we will occasionally encounter individuals who have experienced panic within, as well as outside their homes.

In the discussion of difficult spaces and places that follows, I will tend to resist employing the loaded and potentially problematic distinction between public and private space (see Duncan 1996), and suggest that we can better place and understand agoraphobic concerns in relation to the *sociality* of space.[6] The invasive kind of 'sociality' I have in mind tends to be encountered most consistently beyond the home, in, for example, social (and performative) spaces of consumption, where sheer numbers of others can oppress and confuse, and efforts of design are made to (over)stimulate subjects' senses. Such spaces are, arguably, epitomized by large-scale shopping centres and malls, where stimulants and performances are taken to something of an extreme.

In clinical literature, we regularly find references to and lists of the kinds of spaces and places that agoraphobics will tend to avoid. These predictably often include shops of many kinds, busy streets, restaurants and public transport, and are accompanied by statements about the *rationale* for agoraphobic avoidance. We can summarize and paraphrase such explanations (detailed above) in terms of the agoraphobic's apparent fear of the consequences of panicking in such environments, that is, of being trapped and humiliated in public. I would, however, suggest that we require a more nuanced understanding of the nature and constitution of social space such as that outlined above. There is more at stake here than subjects' anticipation of an aversive and fearful *response*. The agoraphobic's apparent *gut reaction* to social space should rather be seen as an *interaction*, as an embodied, emotional *exchange*, arising from a complexity of meanings, meetings and movements *between* particular selves and spaces. Clearly, this approach moves beyond the search for either internal (self) or external (space) explanations of agoraphobia, and is *open* to relational reconfiguration of both. I would suggest that by thus adopting a receptive, expansive attitude to space as 'performative', we would be better placed to conduct a sympathetic and meaningful exploration of agoraphobic landscapes, those spaces and places that partially 'produce' (and are partially 'produced' by) feelings of panic in susceptible individuals.

In summary, this section has shown that a view of space as 'performative' can help develop geographical understandings of the nature of agoraphobia. By examining the socio-spatial circumstances surrounding experience of panic in particular, geographers can illuminate the growth of dis-ease out of the fluid and shadowy contours between people and places. Clearly, subjectivities and social spaces *interact* to 'produce' disorder, and agoraphobia is particularly well placed to underscore the character and power of this socio-spatial interaction.

Gendered Boundaries

> From the outset urban phobias were assigned a definite place in the gendering of metropolitan psychopathology ... [and] were thought of as fundamentally 'female' in character (Vidler 1993: 35).

Reviews of clinical, experiential, and interpretative literature reveal that the sex of any individual subject is far and away the most reliable predictor for the occurrence of agoraphobia. Thorpe and Burns' (1983) analysis of ten different agoraphobia studies found the average proportion of female sufferers to be 80 per cent. Tian et al. (1990: 317) claim that a '[r]eview of the literature suggests on average 82 per cent of the agoraphobic patients have been female', while Clum and Knowles' (1991) review of twelve different agoraphobia studies puts the figure at 89 per cent.[7] In this section I want to consider why this might be the case. What is it about the gender(ing), situation(s) and treatment of women that entails heightened susceptibility to agoraphobia?

One perhaps obvious place to begin would be with a consideration of gender role stereotypes, which for women, involve characteristics not dissimilar to those found in agoraphobia. As Joy Reeves (1986: 154) points out, a woman's place is 'in the home, and not in the market place' (though it should, however, be noted in advance of this discussion that such 'home-based' approaches to understanding agoraphobia are particular and partial; more boundary focused alternatives will be considered below). Such notions about 'proper' behaviour and roles for women are culturally pervasive to the extent of being clichés; they portray and arguably partially construct women as more dependent, emotional, passive, fearful, home-based etc. (see Moi 1999), a 'fact' not entirely lost on clinicians (Brehony 1983). For example, in response to Clum and Knowles' (1991) finding that men are almost as likely to suffer from panic disorder as women (41 per cent), but are far less likely to go on to develop agoraphobia (11 per cent), McNally (1994: 170) suggests that '[s]tereotypical gender roles may ... foster agoraphobia in those who experience panic'. Mathews et al. (1981) point out that the social role expected of women is consistent with the higher incidence of agoraphobia, and educational psychologist Iris Fodor (1974) suggests that the *key* to the gendered distribution of the disorder must lie with the socialization of girls towards 'helplessness' and 'dependence'. She writes that, in later life, 'women are reinforced for their avoidant behaviour, since it is consistent with the stereotypic feminine role. Since women's place is "in the home", a woman who avoids leaving the home isn't so abhorrent'.

Psychiatrists Dianne Chambless and Jeanne Mason (1986: 231) concur with Fodor's view that sex-role stereotyping predisposes women to agoraphobia, and suggests that women are less likely to be practised at being outside the home and coping on their own. Further, she suggests that because fearful behaviour is incongruous with the male sex-role stereotype, men are more likely to approach a feared object than women, even when both describe themselves as 'equally fearful' (Chambless 1992: 4). Behaving fearfully is, she argues, more acceptable for women and 'a society that does not teach women to be instrumental, competent and assertive is one that breeds phobic women' (Chambless 1986: 234).

Referring to research which attempted to test sex-role stereotyping of agoraphobics, Chambless (1982: 4) reveals that the study 'did not find agoraphobic women to test as excessively feminine, although the males' scores were high on femininity, according to clinical norms'. Chambless also speculates about possible biological explanations, for example, the link between testosterone and dominant, less-fearful behaviour. Unusually, she even casts doubt on the very existence of sex difference in agoraphobia. She writes (1982: 4/5) that

> agoraphobia in men may often be masked by substance abuse. ... Given the prevalence of alcoholism in Western society, it may be expected, then, that large numbers of male agoraphobics are to be found, not in phobia societies and psychiatric clinics for phobia treatment, but in bars, Alcoholics Anonymous, and alcoholism treatment centers.

This study does, as Chambless suggests, warrant careful replication. The fact remains, however, that the overwhelming majority of individuals presenting with, trying to articulate, and requiring help for agoraphobic symptoms are women. It is with this group of subjects that I am concerned in this book.

Clearly, any issue relating to gender, susceptibility to agoraphobia included, can never be considered apart from political/historical context. That is to say, any explanatory account of even individual women's experience requires some understanding of the position of women in patriarchal society. Consequently, in relation to mental health issues, there is a significant tradition of feminist theorists who have explored why women apparently experience problems and 'disorder' to a far greater extent than men (Pugliesi 1992). Feminist theorists of women's lives and health in general often begin by pointing out that definitions of 'normality' are constructed in accordance with an always masculine (rational and autonomous) ideal. Women are thus, *by definition*, always already deviant. When we relate this to issues of *mental* normality and health, we find that definitional reliance on conceptual opposition is, however, more than usually explicit. That is to say, mental health is characterized in terms of mental *ill*-health (Bondi and Burman 2001), which in turn, is characterized in terms of 'feminine' traits and behaviour; the irrational, emotional, and of course typically, hysterical. But, as Joan Busfield

(1996: 118) points out, the links between gender and mental 'health' are more complex.

> [W]hilst gender is embedded in constructs of mental disorder, the constructs of disorder which are developed and elaborated by a range of mental health professionals, incorporating ideas about causation and treatment as well as about symptoms, in turn contribute to the way in which gender is itself constructed (see also Busfield 1984).

Conceptions of feminine delicacy and derangement underpin and feed into each other.

Attempting to summarize the diversity of approaches taken by feminist theorists, Pauline Prior (1999: 78) writes that

> [r]easons given for the overrepresentation of women in psychiatric statistics have included arguments about the intolerable constraints involved in traditional female roles, the acceptability of illness as a mode of protest and attention-seeking for women, and the inability of a male dominated society to accept creative but different female behaviour (See also Showalter 1987; Usher 1991; Astbury 1996 and Shimrat 1997).

More pithily, Busfield (1996: 139) arguably encapsulates feminist concerns when she asks '[d]oes women's oppression drive them mad?' A question somewhat similar in nature has in fact been fairly prominent in my mind throughout this research, shaping its course and influencing my choice of theoretical resources for this book: does women's oppression (in terms of their gender(ing), situation(s), and treatment) drive their phobic retreat from the social world?

Of those few social theorists who have offered gender sensitive insights into agoraphobia specifically (for example Gardner 1984 and Garbowsky 1989), some have attempted to *contextualize* its emergence, occurrence and experience in terms of wider cultural (e.g. Reeves 1986), economic (e.g. Brown 1987) and political (e.g. da Costa Meyer 1996) circumstances. For example, Abram de Swaan's (1981) 'sociogenesis' of agoraphobia provides an account of the historical restrictions on women's movement in public. That there is surprisingly little historical literature on the subject is perhaps, he speculates, because these limitations were considered too unimportant or too self-evident to warrant description, much less discussion. In fact, he claims, the richest source of information concerning the 'gradual disappearance of women from the streets during the first half of the nineteenth century' (de Swann 1981: 362), that is, during the initial phase of industrialization, is to be found in etiquette manuals. The injunctions contained within highlight effectively the extent to which 'public' places had become strictly the preserve of men alone. Joy Reeves also notes that, around the time of the 'discovery' of agoraphobia, women were seriously discouraged from venturing outside the home on their own. However,

> [t]oday we are faced with the paradox of women who were originally prohibited by law and custom from entering public areas of activity and who are now diagnosed as phobic when participation in the public arena makes them unduly anxious (Reeves 1986: 154).

It is her contention, and one which this book supports, that '[t]he cure for agoraphobia may ultimately be sociological, which includes the political dimension' (1986: 157).

Reeves' article is valuable in that it emphasizes the need to view agoraphobia as part of a larger picture. She argues that simply '[t]reating symptoms leads to trivialization of human problems and the people who have them' (1986: 156), and that we therefore need to develop alternative, non-clinical approaches: 'When we understand social structures and structural changes as they bear on more intimate experiences, we are able to understand the causes of individual conduct and feelings of which people in specific milieux are themselves unaware' (Reeves 1986: 153, referencing Mills 1959: 163). Issues of sex inequality and its effect on social institutions are, on this account, particularly relevant. Consequently, in accordance with feminist aims and objectives, putting the beginnings of a 'cure' into place for Reeves would involve the creation of 'real options' for both sexes, in the home as well as the market place, and the wide scale reduction of sex inequality in institutional structures.

One alternative avenue of enquiry I would like to explore here relates less to the socialization and treatment of women than to chronologically (developmentally) earlier questions of boundary formation. Given the centrality of boundaries to the phenomenology of agoraphobia, and the fact that the corrosive aspects of social space seem to impact more severely on women than on men, I want to question whether we might usefully frame our inquiry in terms of the *engendering* of subjects' boundaries. Are, we might ask, women's boundaries different, perhaps *less protective* than those of men?

Perceptive and potentially useful work on boundary issues has been carried out by geographers informed by psychoanalytic theory. Human geographers such as Bondi (1999a), Rose (1993) and Sibley (1995) have looked to the objects relations approach taken by, for example, Klein and Winnicott, to help explain the 'dynamic uncertainties' and 'instabilities' that characterize subjectivity (Rose 1993: 78). Consideration of some of this work can perhaps further our reconceptualization of the embodied self beyond the *limitations* of the Cartesian rational and masculinist ideal. For, as Sibley (1995) points out, conceptualizing the central developmental processes of projection and introjection enables object relations theory to produce a more 'ecological' model of the self, by emphasizing the *constructive* nature of links between selves and spaces.

Object relations theory offers an account of the emergence of social selves and their protective boundaries in environmental and relational context. On this view, the new-born child has no initial sense of itself as separate from its surroundings. Only gradually does it begin to recognize a distinction between what is 'me' and 'not-me', through a process Winnicott describes as 'personalization' (Davis and Wallbridge 1981: 100). The boundary that emerges during this process protects the individual from dissolution into its environs, and its maintenance is crucial for a sense of mental health and well being. However, these boundaries are never fixed once and for all, but must be continually renegotiated in a dynamic interaction between inner self and outer space. How we, as adults, 'manage' this interaction will be influenced by our experience of this initial phase of boundary formation. As

Liz Bondi (1999b: 4) explains, the process of personalization – becoming a person – 'generates a framework that, in one way or another, resonates through all future relationships' (see also Bondi 1999a; Latham 1999), including, one imagines, relations with places as well as persons. Importantly, for object relations theorists, the sex of the child has a central role in this process, and may help explain why 'some people will have a greater boundary consciousness than others' (Sibley 1995: 7).

In her early (1978) account of object relations theory, Nancy Chodorow argues that the very gendered nature of personalization means women tend to have a greater sensitivity to the fragility of their boundaries than men. She explains that boys are required to develop a greater sense of autonomy, separateness and distance from others, as a result of the different gender of the primary carer (still overwhelmingly likely to be a woman) from whom they must distinguish themselves. In contrast,

> because they are parented by a person of the same gender, girls come to experience themselves as less differentiated than boys, as more continuous with and related to the external object-world, and as differently oriented to their inner object-world as well (Chodorow 1978: 167).

Simply put, 'the basic feminine sense of self is connected to the world [while] the basic masculine sense of self is separate' (Chodorow 1990: 119).

Chodorow's theory may go some way towards explaining some aspects of women's increased susceptibility to agoraphobia. For example, she argues that 'women's relational self can be a strength or a pitfall in feminine psychic life: it enables empathy, nurturance, and intimacy but *threatens to undermine autonomy and to dissolve self into others*' (1990: 121, emphasis added), precisely the kind of fluid phenomena feared by agoraphobic women. Feminist psychotherapists Luise Eichenbaum and Susie Orbach (1992: 165), following object relations theory, in fact suggest that agoraphobic women have

> had to create false boundaries because the possibility of genuine separateness was not available to her in her development. A woman may fear that if she steps outside the home she may lose the physical boundaries she has come to rely on and hence she will lose herself.

On this account, women's boundaries simply do not offer them the protection afforded by men's, and it is this vulnerability to (the effects of) one's surrounds that engenders panic and avoidance in the face of social space. While Chodorow, among others (for example Dinnerstein 1978; Gilligan 1982) has been criticized for 'reproducing the familiar structure of binary oppositions in conceptualizations of gender' (Bondi and Burman 2001:13), she has, in more recent work (Chodorow 1994; see also Flax 1990), taken account of the 'multiplicity' and 'ambiguity' of gender.

Williams and Bendelow (1998: 118) are among those who emphasize that there are certain disadvantages to such dualistic approaches, and to conceptualizing women's boundaries negatively, as not simply different from the masculinist 'ideal', but less than, inadequate in comparison. This is one of the problems we encounter in Iris Marion Young's (1990) 'Throwing Like a Girl', a remarkably spatialized account of women's embodiment, but one which, nevertheless, portrays femininity as a 'liability'. Young argues that either through women's 'training', or an awareness of 'objectification' by the male gaze, women are likely to demonstrate styles of bodily comportment that are more restricting and restrictive than those adopted by men. They are unable to 'master' their environment, and more likely to interact with it cautiously, rather than confidently. As women, we are 'physically handicapped' in (or perhaps *by*) a 'sexist society' (Young 1990: 153).

Clearly, it is difficult to think beyond predominant masculine models of embodiment portrayed as the 'norm', when powerful masculinist discourses and practices continue to permeate social and academic life. For example, as Williams and Bendelow (1998: 115) point out

> contemporary medical textbooks continue to portray the male body as the standard against which the female body is judged, and comparative references to female anatomy continue to employ terms such as 'smaller', 'feebler', 'weaker', 'less well developed', to demonstrate how women differ from men.

But, and as I discuss in chapter 6, Young elsewhere attempts to subvert the masculine norm in essays dealing with 'breasted experience' and 'pregnant embodiment' (1990), a 'gynocentric' phenomenological project also taken up by Christine Battersby (1998).

Battersby's explicit concern in *The Phenomenal Woman* is to 'ask what happens to the notion of identity if we treat the embodied female as the norm for models of the self,' (1998: 38) suggesting we might thus subvert the rigidly contained androcentric norm. In a critique allied with much contemporary feminist geography, Battersby (1998: 57) argues that boundaries do not

> need to be conceived in ways that make the identity closed, autonomous or impermeable. We need to think individuality differently, allowing the potentiality for otherness to exist within it, as well as alongside it. We need to theorize agency in terms of patterns of potentiality and flow.[8]

This account, reminiscent of Irigaray's (1985) conception of 'fluid mechanics', could be particularly useful here. Battersby shows, as I myself repeatedly argue in this book, that women simply do not experience themselves, and their 'place' in the world, in the *limited* manner suggested by the dominant and pervasive Cartesian model of subjectivity. Feminist theorists and therapists thus need to create a more inclusive conceptual framework, what Battersby terms a 'revolutionary space' in which more fluid selves and spatial relations can be envisaged, articulated, and perhaps, more comfortably *managed*.

Fluid boundaries need not, then, necessarily be conceived and thus perceived – actually felt and experienced – as deviant and pathological. By rethinking subjectivity *through* the bodies of women, particularly 'disordered' agoraphobic women, we can highlight and (un)fix the limitations currently in place. The point to be emphasized here is that any 'other' experience of the world *cannot be understood* if we stick with a Cartesian view of the individual as fenced off and radically separate from their environs. Embodied and especially *spatial* difference must be able to be *thought*, if it is to be subjectively integrated and 'handled'. However, I want to emphasize that my aim here is in no sense to 'celebrate' 'disordered' subjectivity. Rather, my intention is to sympathize with, and *synthesize* agoraphobic experience within more inclusive conceptions of selves in space.

I want to close this section by emphasizing the importance for geographers of recognizing that experiences of space and place in patriarchal societies are necessarily gendered, and moreover, that this is stressed by experiential accounts of agoraphobics. A study such as this can, therefore, help geographers understand and theorize the ways in which women and men conceptualize and negotiate their place in social space, and so help elucidate the gendered nature of geographical experience in general.

Conclusion

In conclusion, it is hoped that this study will make an original contribution to wider debates in human geography, and connect some existing areas of research on, for example, space and subjectivities, bodily boundaries, and gendered, disabled and 'fearful' experience of space. By conducting in-depth research on some of the more extreme ways in which people, and women in particular, relate to their environments, we might shed some light on some of the more taken for granted aspects of 'normal' (that is, non-pathological) person/place relations. The substantive chapters of this book will thus present a *constructive* and philosophically informed engagement with agoraphobic experiences. This constitutes a partial critique of existing geographical accounts of everyday interaction between people and places, and aims to forge a path through agoraphobia towards more open ways of thinking and treating spatial relations. By thus developing an increasingly nuanced and feminist philosophical geography, the book demonstrates the potential importance of existential phenomenology for fully spatialized understandings of subjectivities, of use for phobics and non-phobics alike. First though, I will present a discussion of the practicalities of actually doing research with agoraphobic subjects, highlighting what was learned about the negotiation of group boundaries through humour.

Notes

[1] There are however exceptions; Thorpe and Burns' (1983) study found 9 per cent of their respondents developed the disorder before age 16, and 13 per cent after age 40. Marks and Gelder's (1966) study suggests there may be two peaks in age of onset, the first in late adolescence, the second around age 30. Most studies have found that agoraphobia in childhood is very rare indeed, though some links have been suggested with school phobia (Bowlby 1973) or separation anxiety in childhood (Brier and Charney 1985). Marks (1987: 328) suggests that '[t]wo compatible explanations are that such childhood anxieties sensitize someone to later agoraphobia, or that both problems reflect a generally fearful disposition'.

[2] In Words, Sartre wrote, 'I was Roquentin; in him I exposed, without self-satisfaction, the web of my life; at the same time I was myself, the elect, the chronicler of hells' (in Moran 2000: 367).

[3] For example, Merleau-Ponty explicitly suggests, like Lacan, that the unconscious is structured like a language. Moran (2000: 432) also outlines the influence of phenomenology on Foucault and Deleuze and reiterates Derrida's views on Merleau-Ponty's intellectual 'brilliance' (2000: 432). While his contribution has been 'almost completely ignored', there are, Moran claims, grounds to argue that Merleau-Ponty is actually the 'father' of post-structuralism.

[4] Although perhaps McDowell (1996: 32) might sometimes be regarded as tending to replace an absolute notion of space with an abstract notion place as an 'absolute location', as somewhere static.

[5] However, for Gregson and Rose (2000: 441), there is no 'just space'; '"stages" do not preexist their performances, waiting in some sense to be mapped out by their performances; rather, specific performances bring these spaces into being'.

[6] Where the term 'public' is used, I refer to spatial situations stereotypically or colloquially conceived in such terms, for example, public parks, public transport, the public eye; generally speaking references to 'public', should be read as designating 'social' space.

[7] Some researchers have pointed to evidence of cultural specificity in the sex distribution of agoraphobia; Barlow (1988: 28) writes that 'women are strikingly more afflicted with anxiety in Western cultures, while males seek treatment in equally remarkable numbers in Eastern cultures'. Seemingly supporting this view, Marks (1987) points to a study by Chambers et al. (1982) which found 78 per cent of agoraphobics attending a psychiatric clinic for treatment in India to be men. Marks does not, however, accept that the findings are representative of actual incidence, suggesting that they are reflective of referral patterns and that it is culturally more acceptable for women to be afraid of public places. (Though the issue of cultural difference is rarely the subject of clinical discussion, Last and Herson (1988) suggest that agoraphobia should be treated exclusively as a culturally bound disorder of Western origin; it is not, they claim, to be found in Botswana.)

[8] In contrast with more extreme poststructuralist views, however, Battersby (1998: 57) sharply rejects calls for 'abandonment, transgression or deconstruction of boundaries'. She does not, contra Williams and Bendelow (1988: 120), conceive boundaries as 'essentially masculine'.

Chapter 2

'Joking Apart...': The Negotiation of Group Boundaries Through Humour

Every joke is a tiny revolution. (Orwell 1968)

Introduction

Following the previous chapter's review of the various bodies of literature deemed relevant to this investigation, this chapter turns to a discussion of the way the research itself was actually conducted. In so doing, it attempts to document some of the ways in which I came to regard the form of qualitative feminist research appropriate to the topic as a *dialectic*, where those being researched affected the *form* as well as the *content* of the 'finished' work. This is what I have termed a 'processual methodology', intended to convey an approach that is flexible, reflexive, and open to change in response to the needs and desires of the subjects of research.

Prior to embarking on the empirical component of the project, I had understood feminist research methodologies to presuppose the researcher's ability to organize their activities in such a way that they purposefully narrow the 'communicative distance' between themselves and their respondents. However, I soon found that I had seriously underestimated the degree to which membership of a self-help group itself could become an important factor in mediating the respondents' sense of self and group identity. The self-help group can offer its members a collective identity based on the mutual recognition of shared experiences and a shared 'form of life'. This collective identity is not only important as a topic for research, in terms of the relations between the group-members themselves. It also has other methodological implications because the 'communicative distance' between researcher and researched becomes brokered through group dynamics. This chapter thus seeks to explain how and why the recognition of self-help groups' importance in respondents' lives triggered changes in my approach to this research. In particular I seek to do this by highlighting the deployment of humour as a key element of the discursive process.

Given the seriousness of agoraphobia as a condition it may seem odd to focus on humour, but there are good reasons for doing so. First, the ability to share a joke can itself act as an indicator of the degree of group-identity, since joking depends upon those involved being able to call upon a tacit knowledge that others may not possess. That is to say, 'getting the joke' is dependent upon one recognizing the relevance of what is being said about the 'object' of that joke (e.g. agoraphobia), making a particular kind of link to that 'object', and so on. Second, in calling upon this tacit knowledge, jokes (often inadvertently) make *explicit* the things that matter to that group, things that might otherwise remain implicit or un-remarked. Jokes can thus provide the researcher with novel topics for reflection that may influence their lines of enquiry. Third, humour can also re-present and indicate 'communicative distance' in another way because it is often employed tactically in ways that concretize or problematize group-identities. Humour can certainly help bind people together but it is also a tool that can be deployed to signify someone's exclusion or distance from a particular group.

Humour is then an element of, and a way of communicating, a (self-help) group-identity. This group-identity provides an alternative locus of discursive power that can, in certain circumstances, operate in an exclusionary fashion. When necessary a joke can be used to put the unwary, inexperienced, or over-confident researcher 'in their place'. At other times the joke can be used in such a way as to include the researcher, to make them party to what is happening. Reflecting upon the way in which jokes are employed thus provides a way of ascertaining the degree to which a researcher has, or has not, successfully reduced the communicative distance between themselves and their respondents.

To summarize, this chapter recounts the *processual* changes in my own research programme in relation to the various ways in which different individuals in different groups deployed humour (understood as a measure of communicative distance) over the course of the project. In particular it tries to show why agoraphobic self-help groups themselves, rather than just agoraphobic individuals, became a necessary focus of study. I conclude by trying to draw out some of the methodological implications of this study as they might impact on feminist research in general and work with self-help groups in particular.

Feminist Methodologies in Theory and Practice

In order to study the meanings of agoraphobia for those women whose lives are (de)limited by its presence, I became involved to varying degrees with a number of sufferers' self-help groups based in Central Scotland. In truth, my initial involvement was intended merely to establish contact with (around 15-20) group members willing to be interviewed on an individual basis. I had not, at this stage, anticipated quite how significant a place these groups occupied in some sufferers' lives, or imagined that 'a *big* part of [my] thesis should be about groups'. It was partially in response to this exhortation from Carron, one of those women whose words (and actions) have informed this research,[1] that I came to realize that the nature and role of self-help groups was an important topic in its own right. It was also an indication that, however definite my ideas might have been about the project's direction, such plans would inevitably be disrupted, resisted, and forced to evolve in different ways by the agoraphobic subjects themselves.

Prior to embarking on the empirical component of this project, my research design could reasonably be characterised as an almost 'text-book' approach. My prior experience as a feminist researcher was purely theoretical, and so I was reliant on the experience and advice of others concerning the translation of feminist theory into feminist practice. Fortunately, recent years have seen an increase in social science debates on such methodological issues, and this trend has been well represented in feminist geography, notably in symposia such as 'Women in the Field' (in *The Professional Geographer* (46,1)) and 'Focus: Feminism as Method' (in *The Canadian Geographer* (37,1)). The contributors to these debates draw on their own research experience in ways that illustrate flaws in 'traditional' methods, with their emphasis on objectivity and distance, developing alternative approaches based around principles of intersubjectivity, reflexivity, and positionality (see for example, McDowell 1992: 406; Moss 1999: 156). The methodology I devised for this project in advance was then based on my interpretation of such attempts to negotiate ethical relations between researcher and researched.

As a feminist researcher I was also aware of the need for a 'critical and reflexive questioning of what the research/researcher hopes to accomplish, why a particular area was chosen, and for whom we are working' (Nast 1994: 58). I had chosen to study agoraphobia – deemed it 'significant for inquiry' – (Christopherson 1989, cited in Rose 1993: 57) because it is a widespread and debilitating condition that overwhelmingly affects women (Clum & Knowles 1991). As indicated in the introduction to this book, the gendered nature of agoraphobia mirrors that of anorexia nervosa (Bordo 1995). But where this latter disorder has been extensively explored by the social sciences and exposed in the mass media, agoraphobia remains an under researched and largely 'hidden' condition (Bordo 1992, 1995; Bruch 1979; Chernin 1981; Garfield 1984). (The attention given by the popular press to eating disorders is doubtless in part due to the celebrity status of many of its sufferers. In contrast, any media figure that became agoraphobic – that is, fearful of public/social space – would be likely, almost by definition, to rapidly lose their celebrity status as anything other than a recluse.[2]) My initial aims were simply to explore the meanings of agoraphobia *for* its sufferers, and in so doing, allow them to voice their own concerns and perhaps also raise public awareness about their condition. Additionally, there were personal reasons for my choice of research interest since for a limited time I had previously experienced some of the symptoms associated with agoraphobia (see below).

Following Maynard's (1994: 17) contention that 'the legitimacy of women's own understanding of their experiences is one of the hallmarks of feminism', I resolved that the 'data' collected for this project should largely consist of contributions from women with personal (embodied) knowledge of agoraphobia, rather than, for example, the (secondary) testimony of non-phobic health care professionals or other (merely) interested parties. In formulating a method for data-collection, I followed the suggestion that 'ethnographic methods, including participant observation and in-depth interviewing, are powerful in allowing us to both describe women's actions and reveal the meaning of these actions *for them*' (Dyck 1993: 53, emphasis added; see also McDowell 1992: 406). I felt that one-to-one interviews would allow for sensitive exploration of difficult issues, and that these should be open-ended and largely unstructured, to allow participants to construct their narratives in the manner, depth and at the length of their own choosing. I wanted to avoid specifying themes, or making assumptions about subjects' life-worlds in advance (Maynard 1994: 13). Consequently, my interview 'schedule' was loosely constructed around broad opening questions (e.g. 'how long ago did you become agoraphobic?') and prompts to encourage, but not compel subjects to provide more detail (e.g. 'could you tell me a bit about what happened at that time?').

Before the research commenced I struggled at length with thoughts of a *communicative* ideal, the possibility of conducting the interviews in a non-hierarchical and non-exploitative way, perhaps even in a manner that would be beneficial or enjoyable, rather than stressful or embarrassing for the women who take part. Encouragingly, I found that some feminist research had suggested that it was at least possible for interviews to be a rewarding experience for respondents. For example, Maynard (1994: 17) claimed that subjects could be personally *empowered* by their participation in a research project via

> their contribution to making visible a social issue, the therapeutic effect of being able to reflect on and re-evaluate their experience as part of the process of being interviewed, and the generally subversive outcome that these first two consequences may generate.

When recruiting interviewees I had also drawn extensively on previous research in the attempt to construct a suitable and *ethical* method of bringing agoraphobic women's stories to light. There are clearly many methodological pitfalls to avoid in feminist research (England 1994) and much about interviews in particular that can be emotionally dangerous for respondents. The 'give' and 'take' of these 'exchanges' can each all too easily become unidirectional, leaving one party enriched, the other exposed and vulnerable (Katz 1994; Kobayashi 1994).

Consequently, I felt that as the material I accessed could be of a deeply personal and highly sensitive nature, there was a risk I might initiate or exacerbate distress by negotiating difficult moments inappropriately. I was also concerned that respondents might be led to experience anxieties after I had left the scene, and therefore considered it essential to ensure that the subjects I recruited had access to an adequate support network to help them manage any traumatic issues that might arise. Thus the most obvious place for the search for interviewees to begin was with agoraphobic 'support' or 'self-help' groups. By approaching local and physically situated groups, rather than mail or telephone based organisations such as (the England based) *No Panic*, I aimed to make absolutely sure that each individual subject would be *immediately* involved with an established and solid source of support and in no sense *isolated*.

Agoraphobia and the Self-Help Group

Many of those who took part in this project had suffered from agoraphobia for a number of years before eventually obtaining a diagnosis. While diagnosis is in no sense equivalent to a cure, feelings of immense relief, of recognition and legitimacy are common among newly diagnosed agoraphobics. For example, when I ask Carron, who developed agoraphobia over thirty years ago while living in Germany on a posting with her husband, if her own diagnosis after almost a decade of suffering was helpful, she replied:

> Oh aye! Well what do *you* think? You're getting all these weird feelings, nobody's telling you what's wrong with you, you think you're going *mental*. If somebody says you've got, puts a name to it, uhuh, you feel, you feel bloody good [laughs] but it doesnae make it better.

Having something nameable gives sufferers some*thing* to work on, a new linguistic category with which to organize their experience, and offers new hope that it is also treatable. To such ameliorative ends, many newly diagnosed individuals are given the opportunity to access support from self-help organizations and groups. Such information is not always provided immediately, at the primary health care level, but there are many ways in which sufferers can, sooner or later, discover the existence of support networks.[3] For the respondents in this study at least, finding a group to attend has been an overwhelmingly positive experience. However, it must again be stressed that my strategy of recruiting respondents through self-help groups effectively screens out the negative responses of those dissatisfied individuals who no longer attend.

Self-help groups perform a number of important services for their members, including the provision of practical help and advice about managing day-to-day tasks. As Fiona, a divorced grandmother, part time care assistant, and organizer of a StressWatch support group explains,

> once you get into a group, and everybody knew what, exactly what you were talking about, you know. Everybody knew, and they were giving me wee ideas … 'why don't you try this or try that?', you know, and it's been great.

Among the services and benefits of the group, perhaps the most obviously valuable is quite simply the *understanding* of others who have personal experience of agoraphobia, and who thus know exactly what its like. 'The biggest thing about self-help groups', according to Carron,

> is they've all been there, they all know what you're talking about … They're no' gonnae ridicule you, they're no' gonnae look at you funny, they're no' gonnae think you're an oddity, cause they all *know*.

Despite the otherwise diverse membership of the groups, their accounts repeatedly reveal a high degree of commonality and mutual recognition among individual members of self-help groups, so much so, that it seems agoraphobia shapes and permeates *most* aspects of sufferers' lives, in very similar ways. The disorder can then be said to constitute a shared and substantive background against which sufferers' lives are lived. Although numerous and significant differences remain between individual sufferers, this agoraphobic 'background' could be characterized, in Wittgensteinian terms, as being constitutive of a common 'form of life'. This idea of a 'form of life' was developed by Wittgenstein to designate in non-essentialist terms 'what it is about a community that makes possible meaning' (Ruder-Baker, 1984: 288).[4] Its precise meaning is difficult to specify[5] precisely because it does not define any essential features that are identical in all cases, but rather points to the myriad ways in which a shared life-world is called upon and given expression in a common language.

Agoraphobia, I would suggest, constitutes an experiential background that means one will *tend* to conceptualize and respond to situations in certain ways rather than others. According to Baker (1984: 278) 'forms of life rest finally on no more than the fact that we agree, find ourselves agreeing, in the ways that we size up and respond to what we encounter'. If one suffers from agoraphobia, an invitation to a wedding, for example, will be greeted with dread, never pleasure, and even lunch at a local café with one or two close friends can be felt a particular kind of torture. Such commonalities of experience enable participants to take a certain amount of mutual understanding for granted in their conversational exchanges. That is to say, they may draw (implicitly or explicitly) on particular and shared (micro)cultural references, and need not continually explain themselves, or 'spell out' their emotions and reactions to the situation being discussed. Precisely because of significant overlaps among their experiences of the world, fellow sufferers will often *simply understand* what is being said in a way that might leave non-sufferers entirely mystified.

That respondents do have shared concerns, despite significant experiential differences, is felt very strongly. Susan says of her own group membership:

> It's very mixed as well, social backgrounds, religious backgrounds, its such a broad spectrum of people. And I mean its interesting as well, cause we've all had such different lives, but we've all got this one thing in common.

This *commonality* of agoraphobic experience is strongly felt, but not easily communicable to 'outsiders', whose non-phobic form(s) of life render them partially deaf, or indifferent, to agoraphobic discursive accounts. Despite sufferers' best efforts, it can feel as though the linguistic means of bridging the experiential gap are non-existent. In one sufferer's phrase, describing agoraphobia to a non-phobic is 'like trying to explain red to a blind person' (Moyra). For this reason, many sufferers would agree with Carron's conclusion. She states:

> You tend no' to talk to, – unless they're *really* interested [laughs, nods toward interviewer] – you tend no' to talk to people about agoraphobia. What's the point? They don't understand it so what's the point?

That group members partially *share* an outlook on the world is evidenced by their ability to share jokes based around their agoraphobic experience, jokes whose meaning would be lost on a non-phobic 'outsider' (as I myself was to discover). Indeed jokes can act as a perfect indicator of belonging to, or being excluded from, a shared community of meaning. If a joke is to work, if it is to be humorous, then we need to be able to 'get it', to understand what it means, to what it refers. Even if you were the kind of person who appreciated surreal humour it would be difficult to find anything funny about an allusion to a *Monty Python* sketch if you hadn't already seen one. In our everyday lives, whether at work, within the family, or through watching television, we become party to, or excluded from, jokes that others share around us. Humour often entails a form of dramatic irony, it is a case of 'being in on it' or 'in the know' (knowledge) about something someone else might not yet be aware of. There are private jokes that might, for example, entail the repetition of a particular phrase that have some special meaning only because of previously shared experiences. Those who have not lived through these experiences, or have not had them recounted to them, will inevitably fail to find them funny, indeed they may not even recognize them as a joke at all.

The joke makes reference to a particular context, often an everyday context, but one that is approached from an unusual angle, or one revealed, through a sudden change in circumstances or perspective, to have alternative meanings. The delivery of the punch line is often a matter of revealing just what this alternative perspective entails or how it came about – the shock of this revelation is the cue for laughter. Not getting the joke is a matter of failing to contextualize it, to see the difference between the everyday and the way the everyday has been transformed through this recontextualization. To take a trite example, I once repeated the following joke in the context of a feminist discussion group.

Q – How many feminists does it take to change a light bulb?
A – That's not funny!

It has to be said, there wasn't much laughter, which some might consider ironic in the extreme. However, I would prefer to think that this was not because, as the joke suggests, feminists lack a sense of humour (especially about ourselves), but because our collective experience of feminism was anything but humourless.

The ways in which we react to jokes are very complex, precisely because the joke, like language itself, can call upon many different aspects of a form of life in many different ways. A joke has the character of 'depth' precisely because it can be likened to a 'misinterpretation' of normal linguistic usage. Such misinterpretations call upon and reveal 'deep disquietitudes; their roots are as deep in us as the forms of our language and their significance is as great as the importance of our language' (Wittgenstein, 1988, #111). By this I certainly don't mean that we should, like Freud (1976), necessarily understand jokes as a form of cathexis, as being associated with the retention and subsequent discharge of pent up psychic forces concentrated upon a disturbing object (in this case the subject's agoraphobia). No one denies that telling a joke can relieve the tension or that being able to laugh about something can be therapeutic in certain circumstances. (There may well be a kind of 'joke-work' analogous to 'dream-work' to use Freudian terminology.[6]) But the 'light bulb' example also points to another frequently experienced element of jokes, namely the fact that they can also be *discomforting* (Powell 1988). I do not simply mean that some jokes can be judged as being in bad taste, but rather that one is sometimes unsure whether or not it is appropriate to laugh. This is especially true when someone seems to be making a joke at their own expense and/or when that joke relies upon drawing attention to some aspect of their lives in which we apparently do not or cannot share. Here the 'meaning' of the joke is ambiguous because it seems to express a set of circumstances that are not our own and may not normally be regarded as a suitable subject for humour. Our flustered reaction is not a matter of failing to get the joke but of being unsure how to react. We are unsure precisely what *message* is being communicated to us. To what degree and how should we respond to the serious subtext contained within and transmitted by the joke? To simply read the joke at face value is to risk laughing *at* and not *with* others and thereby risk further distancing ourselves from them communicatively. In this sense then, I wish to emphasize the 'social' rather than the 'psychoanalytic' aspects of the respondents' employment of humour.

People are, of course, well aware that humour can induce such discomfort and in certain circumstances can deliberately employ this ambiguity to indicate difference or demarcate a boundary between in-groups and out-groups. On my first day of empirical research, the first person who entered the ante-room of the support group to be interviewed was Carron, who immediately states, 'they [the other group members] sent me in first cause they said I'd scare you' [laughs]. Carron then goes on to suggest that I reposition my chair and my tape-recorder. 'If you turn it round this way and bring your chair over a bit it'll be able to hear both of us.' Later, having spent around thirty minutes recounting her tale of agoraphobia with much passion and eloquence, Carron is apparently satisfied with her contribution, at least for the time being. She is also concerned to make way for the next subject who has volunteered to take part, and so brings the 'interview' to a close. 'That's just about all I can tell you ... You'll get a lot fae the other ones too because everybody has a different perception of it.' And on this note, she requests that I keep her abreast of the project, and departs.

From the outset, Carron's tactics enable her to take a degree of control over the situation. Before having met me, she and the other group members had apparently collectively decided that I was most likely of a kind who needed to be 'put in their place', perhaps intimidated a little and so 'brought down to size'. Her opening statement, revealing that Carron could, or even *should* scare me, is said light-heartedly, and is clearly, explicitly, intended as a joke – that it is encompassed by laughter makes this plain. However, there is perhaps *implicitly* a suggestion of another meaning, a serious matter underlying the humorous facade. Carron is thus – to use an old but pithy Scots' phrase – 'half joking, hale [whole] earnest'. And, at the time of her utterance, I did indeed feel that there was more jibe than jest, and understood Carron's joke (as partially intended?) to show that group members were calling (at least some of) the shots. Indeed, Carron's 'performance' as respondent in this interview is characterized by assertiveness throughout, and it appears Carron is not about to let herself be victimized, belittled, or exploited by my research.

At this point, it would be helpful to situate Carron's 'social' style in relation to background information about her life and experiences that emerged in later meetings. Carron, it seems, has developed an assertive, at times almost aggressive conversational manner in partial response to the very difficult and distressing long-term relationship she maintains with her mother-in-law. It is Carron's conviction that the latter believed that she, working class, and a 'clippie' (or, also in Carron's words, 'just' a bus conductress) when they met, was never good enough for her son; 'she didnae like me and she made it quite clear she didnae like me ... she wanted him to marry this librarian'. Carron had, however, previously been confident and outgoing, and this reaction 'didnae bother me ... it just run off my back, like water off a duck's'. It was only later, when she had returned to Scotland with experience of more serious (phobic) attacks on her person, that her mother-in-law's negative attitude really began to take effect. Carron explains:

> When I really got into the phobia, your self esteem is, on the ground, you know it just is, on the ground, and if you've got no self esteem, then anything, everything anybody ever says about you, you know it goes really deep.

Eventually, and awfully, what she perceived as almost constant manipulation and emotional abuse led Carron to believe that, well, 'maybe I am nothing'. However, a somewhat defensive, self-protective sense of pride would never allow Carron to give public expression to such (ontological) insecurities, and, she feels, she developed an increasingly 'hardened' manner to cope with her few social engagements as a result. Carron is clearly aware that her manner can be off-putting, but, as on the occasion of described above, realizes this need not always be a disadvantage, that it can be useful to take control.

At the time of our first meeting, Carron's assertive tone is initially set by the bold manner of her entrance, and the intention to (re)position us both – quite literally – in particular ways. I did indeed feel soundly put in my place, slightly on edge, and any preconceptions I might have had about timid phobic personalities were already being thoroughly undermined. Carron's entrance did in fact remind me of the only joke I can recall concerning agoraphobia, which occurred in an episode of the British television sit-com, *Father Ted*. Speculating as to the reasoning behind the half-crazed Father Jack's prolonged retreat to a cardboard box, nice-but-dim Dougal suggests, 'maybe he's agro-phobic'. 'No way!', says Ted, 'I've never known Father Jack to run away from a fight'.

And so, in a matter of seconds, Carron had successfully challenged the anticipated hierarchical relationship between researcher and researched, and she continued to maintain a degree of control throughout the interview. For example, when I suggest that the interview should be shaped by those issues Carron herself considers important, she responds: 'That's good ... It's much better to come in wi' no questions, like, you take it fae the person that knows cause you'll learn much more that way'. In this way, I am shown that Carron is assessing my approach for its appropriateness; clearly, she and other group members are willing and able to comment when they see room for improvement in the nature or direction of the project. I am firmly positioned as learning at their hands; it is they who are knowledgeable, and who should be respected as such.

It struck me as significant in this first meeting that, while I had been concerned *not* to highlight differences between researcher and researched, that is, not to stress my outsider status, it seemed that this was precisely what Carron intended to do. Her comment (*'they* sent *me* in to scare *you'*) introduced a gap between us, and served to emphasize her closeness with, as opposed to my distance from, the group. This issue recurred in the interviews with the other two members, who repeatedly emphasized their sense of belonging within the group, pointing out similarities between themselves and fellow sufferers. It seemed clear (at least with hindsight) that they felt a shared identity that I was very much apart from, outside of, and was repeatedly told I couldn't possibly comprehend. (To provide one more illustrative example; 'with the best will in the world, unless you've been there yourself, there's no *way* you'll understand'.) It was strongly felt and voiced that one has to have personal experience of agoraphobia in order to appreciate what the disorder *means*, and respondents could only expect this embodied comprehension from other group members.

Following some reflection on these initial individual interviews, I became convinced it would be remiss *not* to alter my data collection strategy to incorporate groups into my research. This position was strengthened by Maria Mies' (1983: 73) contention that

> the emphasis on interviews of individuals at a given time must be shifted towards group discussions ... This collectivization of women's experiences is not only a means of getting more and more diversified information, but it also helps women to overcome their structural isolation in their families and to understand that their individual sufferings have social causes.

I clearly had to respond to the way in which the women constantly demonstrated through their words and actions that their agoraphobic identities should be seen in the context of the group as a whole. The best way to achieve this now seemed to require ethnographic involvement with group sessions since this appeared to offer an appropriate and potentially valuable way to open the research to an inside (if not insider's) view of agoraphobia. In this way the project's processual methodology, its need to respond to circumstances and change tack at different points, was born of necessity.

The Dynamics of Self-help Groups and Communicative Distance

In late 1998, I began attending the sessions of the group (WLMI)[7] to which Carron belongs, and additionally began to visit a second group (WASP). When I first attended WLMI's meetings, members extended a manifestly generous and genuine welcome. However, it was clear I should not mistake my physical presence within the space of the group with what it means to be a 'true' insider. Often, it was members' use of humour that functioned as a reminder of this fact. 'In-jokes' worked as markers of group members' shared experience, and as resistance to the imposition of my influence on the group.

In an early meeting with WLMI, a passing mention by Ruth, the eldest member of her group, of the fact (previously known to all but me) that she had a heart attack during a 'therapeutic' trip to Lourdes some years ago caused much hilarity among *all* group members. To be entirely honest, I was more than a little taken aback by what I perceived to be a very black humorous episode (and, with hindsight, wondered if this indeed might have been at least part of the point). I was unsure how to respond, but felt that laughter on my part might be overstepping the line of acceptable behaviour from a relative stranger. There was clearly a level of intimacy between these women that I was not privy to, a closeness that was emphasized by their ability to share laughter about a dreadful experience without being seen as unsympathetic. For a mere acquaintance to have 'made light' of such tragic circumstances would surely have risked causing offence, and this was not the only 'humorous' episode to occur during my first few meetings when I was unsure how (I was expected) to react. I felt that such ambiguous positionality in relation to 'shared' jokes illuminated my status as neither fully inside nor outside the group (Katz 1994: 72).

During these meetings, and also later, when listening to tape recordings of this group, I have been struck again and again by the predominance of laughter, despite (or perhaps because of?) the fact that the substantial content of group discussion is often of an ostensibly distressing nature. For the 'outsider', it can come as a surprise that group members frequently discuss difficult issues with a remarkably light touch. The following extract, for example, is taken from Ruth's account of a trying period in her marriage. (Ruth's husband has since died, and the group in fact helped her through the difficult period of her bereavement.)

> I got that angry I put my coat on and I said 'right that's *it*, I'm outa here and I'm *not* coming back'. He says, 'thhaat's awwright'. He says tae [their daughter] 'don't worry hen, she'll be back in before ye know it'. [Pause] I only got to the end of the building.

All of the women laugh at this point (no doubt Ruth's contribution is funny because they *understand* what she means) and Carron states 'that's the other thing Joyce, you cannae threaten tae leave your man because he knows you're no' gonnae get very far [laughs]'. Such humorous exchanges are associated with those occasions, relatively rare in day-to-day life outside the group, when agoraphobia sufferers can take pleasure from their *difference* from others, and commonality with each other. Their 'abnormality' is not usually the stuff of laughter and there is a tangible sense of cathartic release in the group's atmosphere at such times. The very 'irrationality' of their fears becomes a source of amusement in itself. Carron's effort to let me in on this joke also indicated that she, and perhaps the others, were becoming more accustomed to and comfortable with my presence. It suggests an expectation that I will understand rather than judge, sympathize with rather than dismiss accounts of their 'disordered' experiences, views and behaviours. As protective barriers are gradually lowered, it becomes increasingly likely that one of the women will help me to share the sentiments of jokes I appear not to 'get'.

One aspect of their lives that was frequently referred to by group members was the in/ability to dispose of rubbish outside the home. This was used as a measure of the severity of agoraphobia on numerous occasions, and was itself often considered 'laughable'. One respondent states, 'I couldn't even step outside the door to empty the bucket'; another, 'the dustbin had to be right at the back door', and a third, 'I couldnae empty the rubbish'. Given the frequency with which this seemingly trivial activity – disposing of household waste – was discussed in individual and group interviews, I was brought to reflect on its possible significance. One can speculate that, for some agoraphobic individuals, the space within their home is perceived to be the only area over which they have almost total control. In stark contrast, almost all space beyond the threshold of their home seems unmanageably complex and unpredictable. Under circumstances such as these, the ability to expel the detritus and disorder of daily life from the space of the home is perhaps attributed with heightened importance. Agoraphobic narratives in fact reveal that the characteristic desire to manage the space of the home is at times taken to something of an extreme, and sufferers' (over)concern with domestic matters can result in an exaggerated expression of a stereotypical, 'housewifely' version of femininity.[8] Humour is then used to linguistically manage what sufferers themselves perceive to be an 'abnormal' state of affairs. For example, Iris, a trained social worker, (shaking her head as if to acknowledge the pointlessness) describes obsessively combing out the fringes of a particular rug before she could be content that her living room was sufficiently tidy. In a similar vein, (ex-) community volunteer Brenda jokes about hovering around visitors with dust-pan and brush, and continually exchanging used with clean ashtrays. They are not alone in their intolerance of perceived disarray, or 'chaos'.

As I became a more regular attendee at these meetings my relation to the group shifted and the women's attitude towards me also seemed to change. On a later occasion, for instance, group members discussed their experience of a local screening programme for breast cancer. The 'mammogram' constitutes an instance of intolerable entrapment for agoraphobics, characterized by Brenda as being 'anchored'. By this, she means, 'pinned down' (in this instance almost literally) and unable to escape.

Brenda - I mean when you're getting a mammogram you're *stuck*
Ruth [to Joyce] – they've got your chist between a machine [laughs, and uses hands to illustrate clamping action on breast]
Carron – so, you get a panic then, you cannae get out the damned thing, what do you do?
Ruth - [laughs] see their faces, you're running about wi' your chist hanging out!

While it was clearly felt that I still required some explanation concerning the source of their shared humour, it was the experience of the mammogram rather than panic they saw fit to elaborate. Knowing I had not undergone the former ordeal they offered additional descriptions and information, clarification they felt was no longer required with regard to the latter. I was, to an increasing extent, expected to apprehend parts of their phobic experience, even if I could not comprehend all of it fully. The women's recognition of, and expectations regarding my growing awareness and understanding had, I felt, shifted my status somewhat closer to that of 'insider'. Consequently, less contextualizing information and explanations of jokes were thought to be necessary.

The fact that I did get the jokes, that I began to think certain things funny in much the same way as the group members also illustrates how humour can operate as an indicator of communicative distance. Sharing jokes entails an ability to call upon a tacit understanding of a shared 'form of life'. A regular target for phobic humour concerns the *tactics* some women have adopted for coping with the outside world, tactics that they themselves perceive to be 'irrational' and therefore laughable. For example, it is fairly common for sufferers to 'shield' themselves with, and hide behind, sunglasses or umbrellas, which help them feel less exposed to scrutiny and intrusion by others. More unusually, however, Ruth feels intolerably vulnerable without her hat, and consequently wears it for protection at all times. 'I wear it to keep my head on [laughs] I know its laughable ... [but] I couldnae go to the bin without it.' Ruth offers no explanation for this statement, but all of the women present, including myself, find it funny and join in her laughter. We do not need to have the joke explained to us, we just find it funny, and in a way that perhaps defies rational explanation. Intriguingly, and appositely, Freud (1960: 102) observes that when we laugh at a joke, 'we do not know what we are laughing at'.

Clearly, it is a rare and supportive space in which one can describe such unusual sartorial habits and receive knowing, never mocking laughter in response. Ordinarily, *even thinking about* the prospect of having a panic attack, losing control and making a spectacle of oneself could drive an agoraphobic to anxious distraction. Here, however, the women clearly enjoy their conversations, and it is difficult to convey the pleasure and sense of sharing they appear to experience. One of the main benefits of group attendance is clearly derived from a sense of shared experience, and being able to discuss difficulties with others who understand. Much in the way of background knowledge can be assumed in conversation with 'fellow sufferers', as opposed to the numerous experiential gaps that have to be filled in when recounting tales of panic to the uninitiated.

There was, however, a contrast between my experience of WLMI group meetings, and those of WASP. Here, attendance was again negotiated through the group organizer, Mary, a middle class agoraphobic in her late fifties, whose difficulties initially stemmed from her fear of vomiting in public. However, as it turned out, the dynamics and demands of individual members meant that my methodological approach evolved differently. The WASP group was by far the more formally organized. Members gathered in a church hall, where sessions were actively structured, largely by Mary, who would direct questions to individuals regarding their experience or 'progress', and facilitate any discussion that subsequently arose. In such circumstances, it seemed appropriate to take a metaphorical back seat, and cause as little disruption to ordinary group proceedings as possible. I tended to see my role here fairly straightforwardly, as that of a 'participant observer', attending meetings with a prominently placed notebook to remind others of my status as researcher, but otherwise keeping a relatively low profile. I tried always to be responsive to signals from individuals that they wanted to know more about my research, my own experience of panic, or to talk about themselves or their problems. This would usually happen before meetings began, or during coffee breaks rather than in the course of the sessions themselves. During this time, I tended to speak only when spoken to, or on the odd occasion when it seemed appropriate.

Before every meeting we had agreed that Mary would explain my presence. She was also asked to remind those present that they could ask me to leave at any time. I had also agreed to leave the meeting ten minutes before the others to enable and encourage discussion about my involvement 'in private'. This could then, if desired, be relayed back to me anonymously via Mary. In practice, to avoid disrupting things, I usually took advantage of natural breaks in the proceedings to make my exit. Here and in many other instances one has to avoid employing hard and fast rules. I had to remain alert and sensitive to the significance of, or changes in, group dynamics or individuals' moods, and to be willing to react accordingly. Given the formal nature of WASP group meetings, these aims were relatively easy to achieve. I felt there was a requirement, perhaps an expectation that I perform 'professionally', and keep some 'distance' between myself and group members. There was never any real challenge to the boundary between (their) insider and (my) outsider status, from either side of the fence. In fact, my attempt to follow respondents' lead throughout the procession of WASP meetings meant our relations evolved in a way that, if anything, *strengthened* the divisions between us.

Having said this it seemed that the WASP group was itself somewhat lacking the strong sense of internal coherence that characterized the WLMI group. This may have been for a number of reasons. Perhaps the more formal organization and the centralized role of the facilitator occasionally restricted horizontal channels for communication between members. People may also have had less in common. Individual membership and attendance also seemed to exhibit a greater degree of variation from session to session and, being centrally located in a major city, this group also drew on people from a wider variety of backgrounds. (The WLMI group had been predominantly working class.) But perhaps more importantly than all these things was the fact that the group was not entirely composed of agoraphobics. During discussions certain members sometimes spoke about personal circumstances in ways that were quite distinct from the concerns of agoraphobic members and seemed to have been indicative of other, quite different, psychological problems. This of course meant that they were not always calling upon a shared 'form of life' and did not comprise such a tight community of shared meanings.

As one might expect this meant that humour did not feature prominently in the group. When jokes were made I found that it usually seemed appropriate to simply smile rather than laugh out loud. This conformed to the behaviour of most others in the group and to the stereotype of professional detachment expected of me. This is not to say that no one ever made jokes with agoraphobic associations. Some jokes revealed exactly the same concerns and behavioural patterns exhibited by members of WLMI, such as the tendency to be excessively house-proud (taken up further in chapter 4). For example, screwing up her face in apparent disgust, working class mother of four Susan states, 'I don't like mess. I have a real problem with mess, dirt, untidiness, mess'. She then goes on to relate her anxieties about visitors 'disrupting' her home, one of whom recently 'made *crumbs. Honestly*, I had to practically sit on my hands to stop myself bringing the Hoover in'. However, the response to such jokes was more restrained and they were few and far between. This was partly no doubt because it took a great deal of confidence in this setting to counter the group's formal structure and 'hold the floor' but also because of a need to explain and rationalize the joke to others *in terms that a non-agoraphobic might understand*. Susan could not rely on the unspoken complicity and tacit knowledge so characteristic of the WLMI group. She states, 'I was striving to be this perfect wife, perfect mother, perfect little home-maker, perfect this, perfect that ... I was so, so anxious, right? Striving for this perfection, all the time'. Speculating herself as to the reasons for her behaviour, Susan turns to the group, 'but that's years of being in the house isn't it? You've got nothing else to do cause you're so frightened of going outside. My way of distracting myself from panicking is *cleaning*, shampooing carpets, anything'.

At the time (and also with hindsight) I recognized that the members of WLMI were challenging and resisting my mentally rehearsed performance of sensitive, feminist research in a very different way from WASP. Whereas my positioning in relation to the latter tended to be more passive than I anticipated (as a somewhat peripheral, 'foreign' body), members of WLMI demanded more active, direct involvement in their meetings. This may seem paradoxical but it is indicative of the way in which the WLMI's close-knit community operated to empower them and enable them to intervene directly in the research process. On the first occasion of my meeting the WLMI group, Brenda set the initial (assertive) tone for the meeting, getting the discussion off the ground by questioning me about my attitude towards her disorder. I had met Brenda just minutes before, and had spent a few moments, following introductions, explaining to the group as a whole (there were six women present on this day) that I wanted to disrupt their usual proceedings as little as possible, to simply 'sit in on' their meeting. Brenda's first response to my contribution was as follows:

Brenda – Can I ask you what your opinion is o' people that have got agoraphobia?
Joyce – My opinion?
Brenda – Aye. What you think. D'you think we're off wur heids [our heads]?

So much for just sitting in! In this way, Brenda explicitly and effectively positions me in relation to herself and the rest of the group, highlighting my separateness from herself and then making an explicit demand upon me. Her statement, 'what do *you* think of *us*', draws attention to my outsider position, and feels in no sense like a deferral to my supposed authority as an academic or researcher (compare Longhurst 1997: 37). Brenda clearly feels that, as an outsider, assumptions cannot reliably be made about my views, nor can my sympathy/empathy be taken for granted. (Brenda's narrative features more prominently in chapter 4.) I have entered into, in a sense intruded on, *her* group, disrupting its safe, predictable and manageable space, and Brenda quite reasonably wants my views brought out into the open. In this way my research is made *answerable* to WLMI's members. This positioning is maintained hereafter by, for example, explicit assessment of my contributions by group members. By responding, for example, 'that's an interesting question, but … ', respondents are able to reshape my contributions, and refocus discussion according to their own interests and priorities rather than my own. While such evaluations are never, in the event, entirely negative, they do function as a reminder that my continued presence is entirely due to group members' tolerance. In fact, as one respondent pointed out, the women are not taking part in the project 'for completely altruistic reasons. People like you can help publicise the thing'.

In effect then, once again my methodological expectations had been confounded and I had been forced to adopt quite different research strategies with each group to achieve the desired ends. This even affected the way in which I chose to collect material for my project. The absence of a leader to actively shape conversations in the WLMI group meant that the women present tended to be equally involved in their discourse. Each could direct questions to the others, myself included, so there was no sense in which I could take a back seat and 'observe' the proceedings. It thus seemed less appropriate to use a notebook in sessions with WLMI, but members consented to have sessions tape-recorded, explaining that 'we're no' a secret society'. While the respondents would at times carry on discussions seemingly regardless of my presence, they could also channel their discussion towards me explicitly, and expect substantial responses and involvement. The situation was therefore far less 'controlled', from my point of view, than my largely predictable meetings with WASP; in effect, I could be called to account at any time.

Conclusions: A Processual Approach to Spaces of Self-help

This chapter has described the ways in which a methodological approach has evolved from its initial textbook character through the need to engage in a more responsive way with those involved in my research. Engaging with the priorities of respondents led to a move away from a focus on individual interviews towards ethnographic involvement with two agoraphobic self-help groups, each of differing form and content. While I performed the relatively passive role of participant observer in one formally organized group (WASP), my manner of involvement in the other's, less formal sessions, was far more active, and would change from week to week. In response to such variable circumstances, I have suggested that one should be willing, at times, to follow respondents' lead. The researcher cannot anticipate and plan for every possible eventuality in the interview situation. At some stage, it seems, you – the researcher – might have to fall back on 'common sense', and hope that you have enough in the way of sensitivity and basic social skills to get both you and your subject through the interview relatively harm-free (Parr 1998). My own experience reveals that, no matter how carefully such issues have been thought out in advance, the subjects of our research may have ideas and plans that do not coincide with, and effectively thwart, our own. Such circumstances demand that we enact what Skeggs (1994: 88) terms a 'counter-hegemonic approach to research', recognizing a need to respond to respondents' (often unpredictable) requirements rather than cleaving to a pre-formulated design, regardless of how carefully it was planned or how suitable it may have seemed in advance.

Throughout the course of this project, my method increasingly came to be seen, in this light, not as a pre-existing *frame*work, but as a 'work in progress', or rather, *process*. That is, my method was neither firmly established and fixed, nor 'progressing' in one direction toward a particular end. Rather, it evolved as a continually shifting set of methods that had to respond to changes in my understanding of agoraphobia, my recognition of respondents' relation to their groups and the dynamics of those groups themselves. As Kim England (1994: 82) observes, 'research is a *process* not just a product. Part of this process involves reflecting on, and learning from past research experiences, being able to re-evaluate our research critically'. One of the things I hope that this chapter has done is outline and defend the suitability of what I would term this 'processual' methodology, particularly for working with groups.

I want to conclude by making some methodological points that may prove relevant to feminist qualitative approaches in general, but may be particularly important to future work with self-help groups. It seems clear that researchers must recognize the very important differences in researching what might be termed 'natural' groups from 'artificially' constructed groups like focus groups (Longhurst 1996). This point is made in a different context by Holbrook (1996: 141), who argues that 'groups with people who already know each other and share a sense of common social identity have different strengths and weaknesses from research with groups of comparative strangers'. Rather than being brought together for research purposes, the participants in my research had themselves actively chosen to belong to support groups to help them make sense of and manage their disordered lives. A self-help group like WLMI, where all of its members might be said to share a 'form of life' that sets them apart from others and who all occupy a (quite literally) *meaning-full* community poses particular methodological problems. It requires us to reassess the degree to which the researcher is actually in control of the research environment and, in particular, reveals how negotiating the communicative distance between researcher and researched is a two-way and uneven process. Where the agenda of the focus group is set prior to the meeting by a researcher who regarded themselves as facilitating and directing discussion, the dynamics of the self-help group are radically different. Not only do the discussions tend to be more diffuse and the issues addressed less predictable but the members of the group themselves actively direct the conversations and regulate (through the use of humour amongst other things) the relations between researcher and researched. (Sometimes it almost appears as if it is the researcher who ends up being researched.)

Self-help groups thus problematize in a rather different way the 'communicative ideal' that is often thought to underlie feminist research methodologies. As outlined above, many feminists have pointed out that the relationship between researcher and researched can never be equal. However, my previous reading had led me to expect that it would necessarily be the researcher who occupied the privileged position. I was thus prepared, as Kobayashi (1994: 73) suggests, to try to demystify the university based research process as far as possible.[9] I expected to be able to simply reject the positivist model of a standoffish objectivity and employ a 'conscious partiality' (Mies 1983: 68) in order to break down communicative barriers. But in the case of WASP I found the formal structure and dynamics made this almost impossible. In the case of WLMI I simply had not bargained for the extent to which membership of a self-help group could change the balance of power in the research process. Its members (at least initially) were largely in control of regulating the communicative distance between us, and of deciding when and how to construct and deconstruct barriers.[10] I was, then, largely unprepared for the possibility that my respondents might not enter the situation in the 'submissive' spirit I had anticipated, that they might rather actively resist my positioning, obstruct (albeit 'wittily') any move toward insider status, and reconstruct the interview situation in particular and unpredictable ways.

This brings us to the role of humour in the context of 'natural' groups. Humour, here as elsewhere, serves a number of ends. First, as is widely recognized, it is a form of coping mechanism, a way of 'making light' of difficulties, and therefore rendering them more manageable.[11] Second, it is a way of affirming shared experience – the women understand each other and jokes emphasize this. Indeed, as psychologist Jerry Palmer (1994: 58) notes, humour is 'a form of ethological integration [that] aids social bonding [and] the creation and preservation of group identity'. Third, and perhaps most relevant to my methodological concerns, jokes serve as a kind of resistance against my 'authority'. Members' 'performance' of humour *articulates* and *protects* agoraphobic subjectivities from outside(r) influence and intrusion.

I have argued that the manner in which humour is employed in all these ways can act as an indicator of the communicative distance between researcher and researched and of the group dynamics itself. This is because, in order to work, humour has to call upon a basis for shared meanings. The degree to which the joke strikes a chord with those within the group is thus an indication of the depth of these shared meanings. Whether the joke's meaning has to be made explicit or can remain entirely implicit is also indicative of the degree of communality within the group and of changes in the researcher's relation to that group. It is in this sense that each joke can represent a 'tiny revolution' in the research process. Thus humour should be recognized as playing a potentially significant part in such processual exploration. We as researchers should take care to respond sensitively to jokes as expressions of something of worth rather than dismissing them as mere humorous 'asides' to the more serious stuff of research. Otherwise, the joke could well be on us.

In the following chapter, I intend to shift from methodological concerns to conceptual issues, that is, to the question of how we might understand the experience of agoraphobia as expressed in the words of agoraphobics themselves. It is perhaps appropriate that chapter 3 begins with a discussion of the phenomenology of the panic attack since this seems to lie at the origin of sufferers' development of agoraphobia and at the centre of their anxieties. These anxieties are addressed through the philosophy of Søren Kierkegaard, generally regarded as the earliest thinker of the existential tradition, a tradition to which this book continually returns. In particular, the next chapter relates the onset of the panic attack to spaces of consumption like the shopping mall and supermarket. These are aspects of everyday 'being-in-the-world' commonly and closely associated with women's roles, and ones that loom large in agoraphobic lives and narratives.

Notes

[1] Carron (this is of course a pseudonym) is just one of those I spoke with who believes that the importance of self-help groups is too frequently over-looked, that they are under valued, and hence under funded: 'Somebody has to realise the worth of self help groups, and *do* something about it'. Carron is one of 17 long-term members of self-help groups in Central Scotland who has taken part in a series of (in her case, three) in-depth individual interviews since 1998. All interviews have been audio-taped and transcribed verbatim with participants' permission. Interviews revealed that participants have divergent social backgrounds and economic status, though all were white, had children, and all but two (one divorced, one widowed) were currently married. Ages of participants ranged between 33 and 79 at the time of interview. Only two men took part, neither of whom attended self-help groups on a regular basis at the time of this research.

[2] There has however been some recent interest in the topic of panic attacks (characteristically associated with the onset of agoraphobia) with several famous sufferers proving willing to discuss their experiences. An interviewee who felt strongly that such publicity would challenge the stigma associated with her own condition passed on a cutting (no date) from the popular Scottish tabloid, the *Daily Record*, 'Denise Welch, who plays Coronation Street landlady Natalie Barnes, and Sarah Lancashire, ex-Coronation Street barmaid, have suffered panic attacks'.

[3] Should the individual's GP be unaware of agoraphobic support groups, the recommendation may come further down the line of treatment via a community psychiatric nurse (CPN), clinical psychologist, counsellor or therapist. Alternatively, sufferers might see a group advertisement in their local newspaper, a poster in the town library, or news may be passed by word of mouth (a friend of a friend etc.). Television talk shows and radio phone-ins occasionally feature discussion of agoraphobia, and provide telephone numbers for national organizations such as No Panic, Pax, and StressWatch, or the Scottish Association for Mental Health, who each pass on details of local points of contact. These information sources are clearly not comprehensive or fail-safe, and there is no way of knowing how many individuals slip through this supportive net. (For more information on self-help resources, see the essays collected in Gartner and Reissman (1984; see esp. Katz); Smith (1980) and Rapping (1996).)

[4] Although commentators on Wittgenstein's work disagree about the meaning of his concept of 'form of life', the interpretation developed by Lynn Rudder Baker seems to fit most comfortably with Wittgenstein's own anti-essentialist philosophy (see Davidson and Smith 1999).

[5] Baker states '[i]t is no more promising to attempt to describe what would constitute a 'form of life' *per se* than to attempt to describe what would constitute a background *per se*' (Baker 1984: 277).

[6] In the context of the group meeting, humour can clearly serve a number of positive ends. Recent psychological and psycho-therapeutic literature has suggested that humour has the capacity to reduce anxiety (Palmer 1994: 58), even that it is the 'opposite' of stress (Martin 1997).

[7] At this stage, when I began to attend this group's sessions, they were meeting on a less formal basis than previously, without a facilitator in a housing estate based community centre equally (if not easily) accessible to all members. (These changes were entirely due to a much-resisted withdrawal of council funding.) The names of both groups (WLMI and WASP) will remain in acronym form to avoid identification.

[8] There is a sense, I would suggest, in which agoraphobic women involved in this project compensate for their inability to manage some (outdoor) aspects of 'doing gender' (Butler 1990; 1993) – for example shopping, family outings etc. – by doing other (indoor) aspects to extremes, and thus performing an exaggerated stereotype of the western 'housewife' (Tseelon 1995). It seems that for some home-oriented women such as these, the 'private' space of their very own domestic sphere comes to equal the 'public' (appraisable) face of their femininity. Home, for the agoraphobic, can be experienced as an expression and even extension of their *self*, and negative judgement by (non-phobic) others is thus considered to be no laughing matter.

[9] One way of achieving this is ironically to emphasize the very limited power and reach of the researcher and research, in terms of its influence beyond the academy.

[10] Though of course other factors may have played a part in this including the fact that I was considerably younger than the majority of members of WLMI and that I had a relatively low academic status as a postgraduate student.

[11] Psychotherapist Marie Adams (2000: 160), for example, suggests that humour can provide 'a means of accepting the painful, a way of absorbing and acknowledging what otherwise might be too terrible to bear ... laughter can also be used to distance ourselves from an experience, to diminish its significance or be used as a displacement tool to distract ourselves from discomfort'. (See also Lemma 2000.)

Chapter 3

Fear and Trembling in the Mall: Kierkegaard and Consumer/Consuming Spaces[1]

Introduction

The preceding two chapters have been concerned to describe exactly why and how this investigation into the meanings and experience of agoraphobia was conducted, and to offer reflections on the experience and implications of the dynamics of the research *process*. What I want to do now is embark on the first stage of in-depth discussion of the project's findings, by tracing certain significant aspects of agoraphobia strand by strand.

The current chapter developed directly out of agoraphobics' accounts of the initial panic attack(s) that seemed to presage the development of their condition. As I listened to these accounts a number of commonalities seemed to emerge, in terms of the descriptions of the panic attack itself, the ensuing anxiety experienced by sufferers, and the environmental contexts of their occurrence. The problem of how to develop a geographical understanding of the descriptions I was offered seemed to coalesce around what might be termed an 'existential' problematic. In other words, it seemed to me that the phenomenology of the panic attack itself, the respondents' experiences of anxiety and the manner in which certain places were associated with such experiences all seemed indicative of what might be termed an 'ontological (existential) insecurity' in social space.

Hopefully, my reasons for making this claim will become apparent in this and following chapters as I extend and develop the ideas outlined below. However, as the introduction has already suggested, one way of understanding agoraphobia is in terms of an anxiety concerning a failure in our (usually unconscious) ability to manage the boundaries of our self-identity, that is, to manage our existence as a self. In these terms the panic attack might be seen as an expression of a 'boundary crisis', a temporary loss of self-identity and self-control, that can be induced by a wide variety of circumstances but is usually precipitated by and/or entails feelings of being overwhelmed in crowded social spaces. The agoraphobic thus comes to associate such spaces with feelings of panic and becomes anxious about the loss of control such spaces might induce, leading them to avoid them whenever possible.

Since its inception existentialism as a philosophical tradition has concerned itself with the anxieties associated with the emergence of self-understanding and the concomitant possibility (certainty) that this self can (will) cease to exist at some point. As I try to show below there is thus a *prima facie* case for linking existential anxiety with the fear of loss of self that is frequently expressed in agoraphobics' discourses. Put simply, my argument is that agoraphobia might be interpreted in terms of the anxiety that arises out of the problematizing of everyday 'Being-in-the-world' in/through experiences of social space.

In this chapter, I embark on an existentialist exploration of agoraphobic Being-in-the-world by giving careful consideration to the significant place of shopping in respondents' everyday lives. The desire to do so results in part from the predominance of the subject in respondents' own narratives about panic and anxiety. However, I also want to suggest that spaces of consumption are emblematic of what we might call 'agoraphobic landscapes', those arenas ostensibly 'productive', and subjectively representative of the most acute anxiety and ardent avoidance. Such sites are thus, it is argued, well placed to provide substantial insights into the geographies of agoraphobic women's fear.

In late twentieth century Western cultures, shopping has become one of *the* organizing principles of our lives. In Jon Goss' (1993: 18) words

> planned retail space is colonizing other privately owned public spaces such as hotels, railway stations, airports, office buildings and hospitals, as shopping has become the dominant mode of contemporary public life.

The relative lack of attention paid to shopping is puzzling; until, that is, we consider the extent to which the activity is seen as gendered (Miller: 1997). As feminist theorists have forcefully demonstrated, for an activity to be feminized is equivalent to being trivialized, denigrated, and often relegated to the 'private sphere' (Bondi 1992).

That shopping is *still* largely seen as a 'feminine' pastime is beyond a doubt, as recent empirical research has demonstrated (Falk & Campbell 1997). Colin Campbell claims that children are socialized to believe that 'shopping is basically part and parcel of the activities that help to define the female role, and especially of that distinctive sub-role of housewife'(1997: 167). This stereotype has some basis in reality, in that, as Nicky Gregson writes, 'it is still overwhelmingly women who shop ... just as much as it is women who form the majority of retail sales workers' (Gregson 1995: 137; also see Moore 1991; Lowe and Crewe 1991). Given the centrality of shopping to the feminine sex-role stereotype, indeed, to women's gendered *identities*, we can imagine extra difficulties for the agoraphobic woman in her exclusion from this symbolically private, but physically very public, realm.

Thus this chapter marks the beginning of my attempt to employ a feminist take on existentialism in order to understand agoraphobia as a gendered condition mediated through modern (consumerist) environments. It draws on interview material in order to focus on those experiences of 'panic' that respondents associate with the initial development of their agoraphobic anxieties. Given that the chapter thus intends to 'begin at the beginning' – of the theoretical interpretation, and at the onset of the agoraphobic condition itself, it is perhaps appropriate that it should also begin by calling on the work of the earliest existentialist – Søren Kierkegaard. In what follows I argue that his development of an (albeit esoteric) existentialist analysis of human experience and emotion is in fact peculiarly apposite to the subject in question, and can help us understand the threatening nature of phobic space.

Kierkegaard and All-Consuming Anxiety

The concept of anxiety (*Angest*) is central to the existential thought of the Danish philosopher Søren Kierkegaard (1813-55). For Kierkegaard anxiety is an ontological characteristic of being human, of human existence. To be, and to be aware of one's being, is also to be aware of one's mortality, of the finiteness of that existence. That is, in being properly human, and in reflecting on our human condition, we are struck by the possibility of our ceasing to exist at any moment and the certainty that one day we will indeed die. This threat to our existence is expressed in anxiety. Anxiety is that which brings to our attention a concern with the *limits* of our existence, with the nothingness that lies beyond the boundary of our life and at the very heart of our being. Anxiety is thus a consequence of an existence that is constantly threatened with ceasing to be, it is a correlate of being in the face of non-being, of *nothing*. Thus anxiety differs from fear, because fear is always fear of some-*thing* – anxiety is a relation to nothing, it cannot be pinned down, it has no object, but is a generalized and pervasive feature of our existence.

Anxiety is thus linked to the limits (boundaries) of our existence, to our self-awareness as a (mortal) being. At the same time it is also a consequence of being able to think beyond those boundaries, of our *freedom* of thought. Without being able to think the possibility of our own deaths there could be no true self-awareness, no awareness of the limits of our being and thus no anxiety. From Kierkegaard's religious perspective anxiety is thus linked to the 'fall' – to the origin of the human condition – to the attainment of knowledge of self and of the mortality that is a co-requisite of that knowledge. In eating from the Tree of Knowledge humanity becomes self-aware and is simultaneously expelled into a world of death and anxiety. Kierkegaard thus recognizes anxiety's essential ambiguity. It presages both the possibilities of freedom and of the 'fall'. '[T]he greatness of anxiety is a prophecy of the greatness of the perfection', (1980: 64) – of the *infinity* and boundlessness of thought. But, for the finite and bounded individual, freedom's infinite possibilities carry untold risks (see also chapter 4).

Perhaps unsurprisingly, given these religious associations, Kierkegaard also offers us a (rather unattractive) gendered perspective on anxiety. For Kierkegaard, anxiety is a temptress that 'disquiets' and 'ensnares' the unwary individual, a 'feminine weakness' that 'is reflected more in Eve than in Adam' (Kierkegaard 1980: 64). Anxiety is a product of woman's more 'sensuous' nature, the 'immediacy' and openness she exhibits to the felt experiences of her bodily being and to her worldly surroundings.[2] And yet, despite, or perhaps because of, its close relation to sex and sexuality, Kierkegaard's misogyny will not allow him to follow his logic to its own conclusion; woman's sensuous proclivities do not, it seems, signify her greater potential for freedom. Rather, they merely indicate that her role is to mediate between the world and *man*-kind; 'only through Eve could Adam be seduced by [the possibilities offered by] the serpent' (Kierkegaard 1980: 66). Women are then, Kierkegaard thinks, both more susceptible to, and less able to cope with, anxiety. Ironically, the very sensitivity of those most permeable to anxiety's effects means that they are in danger of being overwhelmed to such an extent that their freedom of thought and action is compromised. Thus Kierkegaard claims, '[a]nxiety is a feminine weakness in which freedom faints' (Kierkegaard 1980: 61).

Even this brief introduction to Kierkegaard's thought illustrates that there is some potential here for understanding (agoraphobic) anxieties in terms of what we might term the 'boundary conditions' of human existence. He also seems to offer an initial account (however unattractive) of why women's freedom might be compromised though their very sensitivity to anxiety. Of course, as shown in chapter 1, studies actually reveal no appreciable difference in the incidence of disorders associated with intense anxiety between men and women.[3] But the gendered incidence of agoraphobia does indeed seem to suggest that this particular form of anxiety is a predominantly feminine 'weakness' and it certainly does effect women's freedom. There is certainly nothing liberating in the disorientating and disabling repercussions of anxiety for the women who suffer its almost constant presence.

Thus, while Kierkegaard's 'conceptualization of anxiety' in this esoteric and misogynist form is far from ideal for a feminist geography, it might still prove a useful starting point for analysis. Kierkegaard not only links together conceptions of the self and anxiety, but explicitly does so in terms of a spatial metaphorics, a language of limits and boundaries. In what follows I want to suggest that this language might express something of women's experiences of anxiety as well as describing its practical (geographic) effects. In other words, agoraphobia does not only result in a gendered 'geography of exclusion' (Sibley 1995) that confines women to their homes. It is also, as I argue throughout this book, a form of anxiety best expressed in a spatial terminology. In this I agree with Marilyn Silverman's existential claim that

> [s]patial metaphors are well suited to capture the phenomenology of the particular form of panic that occurs when the most fundamental sense of existence and connection is at stake (Silverman in Bordo et al. 1998: 89/90).

However, I intend to go further than Silverman and argue that the relationship between space and agoraphobia is not merely 'metaphorical'. The boundary conditions of existence, between 'internal' and 'external' space and between being and non-being are very real and constantly re-worked. As chapter 1 argued, 'space is also a doing ... the articulation of relational performances' (Rose 1999: 248) and agoraphobia seems to express an anxiety about one's ability to regulate and control the performative dialectic between permanence and impermanence, permeability and imperviousness. Accounts of agoraphobic anxiety suggest that it seems to be spatially *mediated*, agoraphobia is *in essence* an im/mediately spatial affair. The non-locatable fear (that anxiety not of any-thing) which typifies agoraphobia, that oppressive, yet inherently intangible 'something in the air', is, I claim, a quality of the dynamic relationship between *self and space, person and place*. For the agoraphobic, anxiety becomes *all-encompassing*, (overwhelming) affecting those spaces in, of, and around the self.

Kierkegaard's phenomenal psychology too recognizes similar requirements for a balanced self which has to be seen not as a 'given', atomic and autonomous individuality but a 'process of becoming' (1989: 60), a 'project that occurs in time' (Hannay 1989: 23). 'A human being is', he says, 'a synthesis of the infinite and the finite, of the temporal and the eternal, of freedom and necessity' (1989: 43). For Kierkegaard, '[p]ersonhood is a synthesis of possibility and necessity. Its manner is therefore like breathing (respiration)' (1989: 70) and, like breathing, it usually continues unnoticed and unremarked unless problematized. The problem is that, once panic has drawn the agoraphobic's attention to the unstable morphology of the self, she then becomes focused wholly on monitoring the self in the face of existential anxiety. As one respondent, Susan, remarked 'I think people with this disorder are very, very selfish. You don't mean to be selfish, but you're constantly tuned into your body, your self, all about you, all the time'. Anxiety becomes all-consuming.

Kierkegaard invokes a spatial metonymics that echoes sufferer's accounts of agoraphobic anxiety. For when the self looks 'down into its own possibility' (1980: 61) its anxiety increases, 'phantasms succeed one another with such speed that it seems as though everything were possible, and this is the very moment the individual himself [sic] has finally become nothing but an atmospheric illusion' (1989: 66). The world begins to spin and the individual must grab 'hold of finiteness to support itself' (1980: 61). As we saw in chapter 1, agoraphobia had, prior to Westphal's 1871 introduction of the term, been referred to by Benedikt as *platzschwindel* – literally 'dizziness in public places' (Mathews et al. 1981: 1). Thus Maggie describes the onset of a panic attack:

> Your heart starts thumping, you feel you're choking in your throat, you can't swallow and you feel all dizzy and giddy, and like if it's in a shop, or the church or what, you just feel you want to run.

For Kierkegaard too '[a]nxiety may be compared with dizziness' (1980: 61). 'He [sic] whose eye happens to look down into the yawning abyss becomes dizzy' (1980: 61). This 'abyss' is not so much a geographic feature of the landscape of one's everyday life as a rift in its fabric. Nor is anxiety simply an 'internal' reflex to specific external stimuli. Rather anxiety is evoked by an existential challenge to what for most of us, most of the time, is a well-defined presumption of our concrete self-identity. 'I couldn't control it when I was out there? You know, really, you honestly believe you are *on the brink* of death' (Susan, emphasis added).

'Dizziness' is, if anything, too mild a term for the intensely disorientating personal experience of the sufferer as she seems to 'lose touch' with her surroundings, to stand at the edge of the abyss. The sufferer feels isolated from all external assistance, sucked into a vacuum that threatens to engulf her very being in an overwhelming panic.

> When you have this you feel you're on your own, you feel [pause] nobody else has got it, you're in this situation, and it's just like a situation you'll never get out of, *a black hole that you'll never get out of* [Ann, emphasis added] .

'But what', Kierkegaard (1980: 61) asks, 'is the reason for this' dizziness? Where does this feeling *come* from? The source of anxiety is not easily *located*. 'It is just as much in his [sic] own eyes as in the abyss, for suppose he had not looked down?' The panic attack comes just as much from within as without and its unpredictability may be one of its most terrifying features:

> Often they come completely out of the blue. A lot of people who come [to the group] say 'I was just perfectly fine, and I was walking down the road, and then, all of a sudden, you know, the pounding and the racing heart, and the sweating' (Mary) .

An everyday situation can suddenly lose its familiarity as the sufferer is caught in a vortex that spins her out of control. Silverman's description of the panic attack is revealing. 'Suddenly the stage disappears. *The floor drops out.* The players and set vanish or persist as unfamiliar figures in another script in which you have no part' (Silverman in Bordo et al. 1998: 90, emphasis added). In such circumstances the sufferer longs for something solid to grab hold of. Susan Bordo describes her own agoraphobic experience thus, 'I was stranded *on the edge* of an enormous [iceberg], rising *high* out of the sea, *perched, precarious, desperate for walls to plant my hands against*' (Bordo in Bordo et al.1998: 80 emphasis added). This too can be described in Kierkegaard's terms. Since he believes that our freedom to act as individuals depends upon imagination – 'the medium of infinitization' (1989: 60), the danger is that

> when emotion becomes fantastic in this way, the self is simply more and more *volatilized* and eventually becomes a kind of *abstract sensitivity* which inhumanly belongs to no human ... the person whose emotions have become fantastic ... *becomes infinitized ... he* [sic] *loses himself more and more* (1989: 61, emphasis added).

This description is very close indeed to the phenomenon recognized in the clinical literature as 'derealization' and/or depersonalization associated with panic attacks involving disturbing and sometimes bizarre changes in one's feelings of embodiment (Chambless & Goldstein 1982, Marks 1987). Roger Caillois refers to this problematization of the individual's boundaries as 'depersonalization by assimilation to space' (Caillois, cited in Grosz 1994: 47). According to Isaac Marks, '[d]uring depersonalization one feels temporarily strange, unreal, *disembodied*, cut off or far away from immediate surroundings' (Marks 1987: 342, emphasis added).

This sensitivity to the infinite, the unbounded, is abstract precisely because the self seems to have nothing solid to latch onto; 'the object of anxiety is a nothing'

(Kierkegaard 1980: 77). And yet precisely because that anxiety is non-locatable it attains a terrible reality. 'Its fear of *fear*. Its fear of what [pause] this feeling that's gonnae come. It absolutely petrifies you, and you just don't want to face [pause] because it's a terrible feeling' (Carron). This 'nothing' is the abyss above which the self, made conscious of its possibilities, balances, fearing to look down lest it should fall and be annihilated. Yet ironically, this nothing is also the open landscape against which the self can choose to define its own existence. It is for this reason that Kierkegaard claims *'anxiety is the dizziness of freedom'* (Kierkegaard 1980: 61, emphasis added).

Consumer/Consuming Culture

> The biggest danger, that of losing oneself, can pass off in the world as quietly as if it were nothing; every other loss, an arm, a leg, five dollars, a wife, etc. is bound to be noticed (Kierkegaard 1989: 1).

For Kierkegaard there are two ways of losing oneself, the first in the swirling maelstrom of the imagination's unbounded possibilities, the second by never becoming aware of oneself at all, for example, by focusing one's attention on *being someone else*.[4] The former is, for Kierkegaard, preferable since it is more reflexive; at least the sufferer has become aware of the pliant morphology of their existence, of freedom's possibilities and dangers. In the second case, the individual may often appear untroubled, though this is a result of their having suppressed their own potential by continually misrecognizing who or what they might be. This he refers to as an impotent *'consumption of the self'* (1989: 48).[5] The fact that the person seems untroubled is, according to Kierkegaard, merely a sign of 'a sickness that [...] hasn't yet declared itself' (1989: 50).

Kierkegaard's reference to the 'consumption of the self' is doubly pertinent given agoraphobics' constant references to consumer spaces – shops, streets, supermarkets, and especially shopping malls – as sites associated with panic attacks. Indeed, perhaps we might argue that the subject who ventures into the shopping mall or supermarket risks *losing herself* in both of the ways Kierkegaard identifies. On the one hand, faced with so many disparate versions of what the self is or could be, she directs herself towards something she is not; she is compelled to seek an identity behind the idealized images modelled in window displays and advertisements. But, even if she resists the imperative to consume subjectivities, to be an 'identity shopper' (see Langman 1994, and Gabriel & Lang 1995: ch.5), she must still negotiate those avenues of corrosive space which seek to break down her resistance to buy, to manipulate her values and emotions, and to forget herself in the momentary pleasures of satisfying desires manufactured by others.

Spaces of consumption present grave dangers to the identities of phobic and non-phobic subjects alike. They are assemblages of numerous and overlaid attempts at sensory stimulation, which can render the atmosphere excruciatingly intense. Susan Bordo describes a panic attack in a crowded supermarket, when everything feels alien and frightening to her, 'the noise, the crying children, the pushing and shoving' (1998: 84). Clearly, this effect is not accidental; the charged space is *intended* to have a certain impact on the subject. From market stalls to mega-malls, shopping sites are deliberately organized to encourage us to buy, to weaken any resolve we have not to. A great deal of energy, not to mention finance, is expended in attempts to pinpoint vulnerabilities to be exploited, and ways to elicit particular consumer responses. This kind of manipulation is indubitably more *intense* in larger shopping centres and malls where sociability takes second place to sales. Whilst every *agora* (market place) has characteristic sights, smells, sounds, tastes, and objects to be picked up and handled, the mall is arguably more single-minded in its orientation toward consumption (see Goss 1992, 1993; also Kowinski 1985; Shields 1989; 1984; Hopkins 1990).

This kind of analysis of shopping spaces is not new. The attempt to draw connections between spaces of consumption and the deconstruction and reconstruction of self-identities has a pedigree that goes back at least as far as Walter Benjamin's *Paris Arcades* project. Burgin explains how, according to Benjamin, the arcades represent an image of space latent in all of us:

> the pre-Oedipal, maternal, space ... In this space it is not simply that the boundaries are 'porous', but that the subject itself is soluble. This space is the source of bliss and of terror, of the 'oceanic' feeling, and of the feeling of coming apart; just as it is at the origin of feelings of being invaded, overwhelmed, suffocated (Burgin 1996: 155).

Shopping in the mall can thus be said to be *contradictory*; 'it is an experience that yields both pleasure and anxiety, a "delightful experience" that can quickly become a "nightmare"'(Falk & Campbell 1997: 12). It is also clearly an experience that problematizes the boundaries of self-identity.

Recent sociological approaches concur with Burgin's psychoanalytic reading in recognizing the impact of spaces of consumption upon self-identity. For example, Bauman describes the subject of consumption as someone who is engaged in 'self construction by a process of acquiring commodities of distinction and difference' (1988: 808), that is to say, buying an identity. We *interact* with consumer products, as part of what has been called the reflexive project of the self, asking questions of ourselves such as, '"Am I like that?"; "Could that be (part of) me?"' (Falk and Campbell 1997: 4). For these reasons the mall is, I would argue, the site of boundary crises *par excellence* as we try to build a secure identity in an environment which continually erodes it.

In this sense at least the shopping mall can be regarded as both *symptom* and *cause* of the loss of stable social anchorings. Not only does the

> mall-based allocation of goods and dreams, gratifications and identities, [provide] no more than intermittent palliatives for underlying anxiety and appropriation of, if annihilation of subjectivity (Langman 1992: 67)

it is actually complicit in developing an architecture of abjection. In the space of the mall, one is subject to more than mere 'sensory stimulation', there is rather a continual perceptual bombardment which, bizarrely, is intended to simultaneously stimulate and sedate. Crawford claims that 'the mall presents a *dizzying* spectacle of attractions and diversions' (Crawford 1991: 3, emphasis added; see also Williamson 1992) that can bring on an attack of *mal de mall*, a 'perceptual paradox' characterized by 'disorientation, anxiety and apathy' (1994: 14). The mall is a place where 'confusion proliferates at every level' (1994: 4). In it, we find

> dream-like illusions and spectacles, eclecticism and mixed codes, which induce the public to flow past a multiplicity of cultural vocabularies which provide no opportunity for distanciation (de-distanciation) and encourage a sense of immediacy, instantiation, emotional de-control and childlike wonder (Featherstone 1996: 103).

The consuming subject is channeled in a certain direction, passing a bizarre mix of design features which have been extricated from any historical or geographical context. For Goss, the mall is 'effectively a *pseudoplace* which works through spatial strategies of dissemblance and duplicity' (1993: 19). The spatial and symbolic architecture of the mall is conducive to the production of agoraphobic-like anxieties precisely because it is an extreme exemplar of that 'amusement society' which, as Langman writes, both 'fosters panic and provides commodified means for its alleviation in ever more consumption' (1994: 68).

The mall in particular, and the shopping experience in general, simultaneously breaks down boundaries and offers a way of re-fashioning them, namely, by buying the consumer goods with which we identify and which stamp a transitory identity on us.[6] Given the close relations this book outlines between agoraphobia and conceptions of bounded identity, it is not surprising that agoraphobics should be especially susceptible to detrimental effects of the shopping experience. Its contradictions and juxtapositions render the space incomparably intense, and are *meant* to problematize our boundaries; its space is so thoroughly charged as to be incomparably *corrosive* to the secure, as well as the agoraphobic, sense of self. As Jane notes; 'it's very stressful for *anyone* [pause] no' just people that have [pause] agoraphobia'.

The relationship between places of consumption, especially supermarkets and shopping malls, and agoraphobia was a constant theme of discussion at meetings and in interviews. In one discussion group, sufferers had an enthusiastic, if inconclusive conversation about the puzzling nature of their extreme, and shared difficulties with 'the Gyle' (an Edinburgh mall they had all attempted to negotiate):

> Ruth – I was only in ten minutes!
> Jane – If I go into the Gyle, even if I go in with someone with me, I start getting this [pause] I want to go home.
> Lizzy – How do we all get it in the Gyle?

They go on to discuss various possibilities, including lack of exits, shape and layout, and the effect of different types of lighting. They cannot, though, put their finger on why exactly it horrifies them so much, and struggle to articulate the nature of their anxieties. I would argue that the apparent intangibility of their shared experiences can only be explained by taking into account the ambience and effects of the 'mall experience' as a whole on subjectivity. Clearly, this must include considering the manner in which the mall is intentionally constructed, so as to over-determine the breakdown of consumer resistance, to undermine the boundaries of the self.[7]

For example, agoraphobic subjects clearly derive comfort from the knowledge that they are free to move as they please, and *leave* if they need to. This requires careful positioning relative to an escape route at all times. Spaces of consumption, however, do not allow for easy enactment of this defence. Even if one succeeds in negotiating the lines of 'goods', it is still necessary to stand in line and present one's purchases at the till before reaching the exit – 'there's a queue in the shop and you think, I must get out of here, I just can't stand in that queue' (Lynn).[8]

Shopping seems to wrest control from the agoraphobic at all levels and, for the subject of uncertain boundaries, shopping is a risky business, best avoided. Avoidance may however be particularly difficult for those with roles as 'wife' and 'mother' and, since it is still culturally coded as a feminine pastime (see for example Campbell 1997; Langman 1994 and Miller 1997), shopping is top of respondents' lists of 'things you should be doing'. In this case shopping is not only a practical necessity but a moral 'duty' inextricably linked with their gendered identity. The inability to perform this duty may be perceived as a disturbing personal failure and initiate serious and debilitating doubts about their self worth.

> One thing I will say, it destroys your confidence. It *takes* your confidence, I mean it really does, it knocks it right out you ... it really screws you up (Susan).

> You feel like a wee ball of nothing. You wonder what you're on the planet for, when you can't do *any* of the things you *should* be doing (Carron).

The agoraphobic subject suffers from her inability to shop in more ways than one. She can neither 'do' the shopping (for necessities, i.e. her 'job') nor 'go' shopping (for pleasure, or contact with friends) and the inability to perform these basic tasks can be very demoralizing, depriving women of the opportunity to reconstruct their identities in crucial ways. As the following quotations illustrate, they feel deprived of the chance to lead 'normal' lives, in a society where predominant ideologies dictate that to be normal is to be a consumer (but see Dyck 1995 and Park et al. 1998 for critiques of such ableist assumptions).

> Your family are suffering. You're doing nothing that a woman [pause] a mother and a wife should do (Carron).

> You feel like, you know, a bad parent, because you couldn't do all the things that the other parents were doing ... I just felt that I wasn't doing my so-called-job properly (Susan).

> I'd love to just meet my daughter on a Saturday afternoon, and just have lunch, and go shopping. I can't do it (Nora, group discussion).

This part of the sufferer's 'job' as a wife and mother, which requires her to go outside the home, is singularly the most difficult activity for her even to contemplate. When it comes to shopping, for example for food, or clothes, or even birthday presents for the kids, the agoraphobic woman has a real problem on her hands, one that almost certainly requires some ingenuity, and outside help, to cope with.

> So then I sort of, if I was going to the shops I would wait for my husband so that he would come with me, you know I found all these different ways. Or I would get somebody else to go to the shops for me, so that basically I didn't have to go to the shops (Susan).

Other interviewees recount a similar kind of dependence on various friends and family members. Fran, for example, describes her mother as having to 'stand in for her'; in Lynn's case 'the kids had to go to the wee shop across the road', or her sister (who also had some experience of panic episodes) would help out. Eventually, Lynn's sister started to encourage her to leave the house herself ('you can't live the rest of your life looking at four walls'), first to visit her at home, two bus stops away, then to go on shopping trips together. 'The first day I got the bus down to her, I was hyperventilating when I got off the bus and it was a *nightmare* [pause] but I made it'. Very gradually, Lynn was able to build enough confidence to try her hand at shopping; 'although I felt uncomfortable ... you don't *enjoy* shopping, you don't enjoy these kind of things, but she was with me and she helped me'.

It is in recounting their experiences of 'consumer affairs' that the respondents reveal most about the effect of the disorder on their everyday lives.

> Non-phobics take a lot for granted. I did too when I was non-phobic. You take it for granted that you just go out and get on a bus and you go. I mean I could never take for granted that I'm going to lift my shopping bag and go and get some shopping. I cannae do it (Carron).

Thus 'shopping' provided a recurring, even unifying theme to their narratives that instantiated both the moment of the first panic attack – their existential crisis – and in their 'geographies of exclusion' – their social and physical confinement. It also recurs in other contexts, for example in Maggie's case, her ability to cope with shopping is an indicator of her degree of recovery.

> So I gradually started going into [nearest large town] with my husband. And I'm no saying it was easy, sometimes I was sitting away down on the floor in the back [of the car] [pause] things like that. But then gradually it did get better and I was staying longer in the shops.

Maggie can then be seen to measure the extent of her recovery by her ability to stay in shops for increasing amounts of time. Even as her tolerance builds and her defences grow stronger, the smallest details of her grocery requirements continue to be of tremendous significance; the replenishment of the larder, even the breadbin, will never again be taken for granted.

I can go the shopping an' that now, and I go down the street every day for the wee [Pause] I do a big shop once a week, and then I have to go down through the week if I need fresh bread or milk or whatever. And I can do that (Maggie).

For Carron, too, attempts to develop spatial strategies which enabled her to cope on a day to day basis, and to regain some control over the geography of her life, were communicated around the theme of shopping. She was put in contact with a community psychiatric nurse (CPN) who helped her to cultivate a tolerance of consuming places. By coming to her home, and after careful preparation, taking her out, then into a shop, then a shopping centre, she gradually and gently increased the time spent there on each occasion. Having prior knowledge of exactly how and when she would leave, a degree of control over the proceedings, made the experience (barely) tolerable. Forewarned, and tentatively forearmed, sufferers can just about manage to contain the rising tide of terror until it is time to make their escape.

Sharing accounts of their problems with shopping can also be seen to form a frequent focus for the self-help groups.

Just the knowledge that other people suffer the way you do is so [pause] helpful. When I'm in the supermarket and I start having a panic attack, and I think 'oh oh', and then I think, 'M will have had these bad turns too', and it just makes me feel so much better (Fiona).

This is one of the reasons why attending the self-help group can be so very constructive. As we saw in chapter 2, the sufferer learns that they are not alone in the world. The other members know exactly how difficult it is to place yourself anywhere *near* shopping spaces, and will give you credit for your painful efforts.

We can give support, and understand [pause] if you do manage to do such a simple thing ... We can rejoice with them. But if you don't know what its like and what its cost that person, to pluck up the courage to do it [laughs], you know if you tell someone that isn't a sufferer, they think, you know, 'well, big deal' (Ann).

However, it must be stressed that for many agoraphobics, *talking* about this issue is all that they can do; actually *going* shopping, *at all*, is simply not an option. When the women I interviewed move on to discuss their experiences of gradually regaining entry to the forbidding, *consuming* realm of shopping spaces, it is in many cases only after having spent *years* in the attempt. Only when they determine to set out on the road to 'recovery' can they begin to think about being a consumer.

Conclusion

To enter the mall without 'fear and trembling' one must become *sure of oneself* – of the self that *you* are – rather than those myriad others that are offered as alternatives. One must also feel secure *in* oneself, at ease in the body's bounds. However, the shopping mall epitomizes those architectures of alienation that undermine security, seeking to play upon the feminine imagination *as consumer* and, at the same time, realize the misogynist *imagery of the woman as that which is consumed* by/for others. The woman who dares to enter spaces of consumption is forced to accept a subject position structured by these two poles, to abandon hope of being herself. Inside the new churches of consumer capital the feminine self is sacrificed according to a patriarchal logic which (in Kierkegaard's words) regards '[a] woman who is happy without *self-abandonment*, that is, without giving all of her self, no matter what she gives it to, ... [as] altogether unfeminine' (1989: 81, emphasis added).

But perhaps hope is not entirely lost. As the respondents in this study show, though they may never become *carefree* consumers, neither must they remain confined within the four walls of their personal domestic realm. It is possible for a path to be forged between these poles, for the agoraphobic to re-constitute herself as a bounded individual in the face of infinite possibilities and the finality of non-existence, who can avoid the 'consumption of the self'. In part this might depend upon coming to recognize the paradoxical nature of the relationship between anxiety, freedom and selfhood.

This paradox is at the heart of Kierkegaard's philosophy. To be aware of oneself is to be aware of the self's possibilities, of its freedom to be otherwise. It is this freedom that makes us autonomous (human) rather than automatons. Through self-reflexivity we become aware of our-selves *as* selves, as a synthesis of the finite (a bounded and embodied subject) and the infinite (freedom), of 'actuality' and the 'potentiality'. But this synthesis is difficult to achieve, for freedom can be a terrifying prospect and looking upon it we are subject to that supposedly 'feminine' weakness, 'anxiety'.

Never having the potential to be a self in her own right, 'woman' seems destined to remain in anxious purgatory. But, Kierkegaard's misogynistic exegesis of the self entirely ignores the feminine predicament and in so doing he misrecognizes both the etiology of and the potential cure for such existential anxiety. Kierkegaard forgets that sensitivity to the self also entails sensitivity to the self's surroundings. *Sensitivity* to the self's possibilities also 'brings home' the manner in which the external world threatens to intrude upon and call that self into question. In other words anxiety is not simply '*the dizziness of freedom*' (Kierkegaard 1980: 61, emphasis added). Anxiety can also be a mark of the self's recognition of those external forces which threaten to subvert its autonomy, to restrict or undermine its hard-fought freedoms. The struggle to maintain a conception of one's *self*-identity is made harder for a woman by her recognition that, in a masculine economy, her surroundings selectively seek to isolate and/or dissipate her identity, that they constantly call her to deny her-self. Indeed the feminine sensitivity to her surroundings which Kierkegaard makes so much of might itself be a product of this recognition of an environment that treats her as consumer and/or seeks to consume her.

If this is true, then there lies the potential for a reflexive recognition of the self, despite the manner in which the masculine economy mediates and distorts the feminine dialectic between freedom and anxiety. In part at least this might come from an understanding that the same phenomena that previously seemed to call her being into question can also serve to enhance her feeling of embodied identity. Susan Bordo writes:

> Being outside, which when I was agoraphobic had left me feeling substanceless, a medium through which body, breath and world would rush, squeezing my heart and dotting my vision, now gave me definition, body, focused my gaze (Bordo et al. 1998: 83).

The feelings associated with panic can begin to be translated as excitement. Like the child who spins in circles, manufacturing dizzyness for the delightful disorientation it creates, the agoraphobic can learn that temporary loss of one's 'normal' perspective can be liberating. To open oneself up to excitement, to learn to endure and even enjoy the *potentiality* of panic without giving oneself over to it completely is the phenomenal freedom to which the agoraphobic aspires. (Indeed, as we will see in chapter 7, the ability to let go in the face of potentially threatening experiences is one of the aims of certain therapeutic strategies.)

However, this process of *re-definition* is slow and often arduous. It is usually only achievable via a process of *acclimatization* to threatening space, and the patience to persevere through almost imperceptible improvements. Very gradually, her feelings of isolation and abandonment may become transformed and even replaced by a sense of freedom and independence. Social space that challenged her boundaries can begin to be experienced positively, and need not entail a descent into the void. Bordo explains how, when overcoming her agoraphobia, she began to enjoy her 'freedom', the

> charge of leaving home, knowing that your body has been cut loose from the cycling habits of the domestic domain and is now moving unrooted across time and space, always to something new, alert to the defining gaze of strangers (Bordo et al. 1998: 81).

Similarly, Maggie describes her elation after having successfully negotiated a trip to hospital to visit her mother: 'Once I got back in [her home town] on that bus, I felt like shouting in the square "I've done it!", you know you just feel so good'. Anxiety, despite, or perhaps because of, its association with the abyss, can be intensely life-affirming, rather than simply soul-destroying.

In the chapter that follows I want to extend the critique of the existential approach developed here, an approach that has been overly entangled in that religious and individualistic perspective characteristic of Kierkegaard's faith. Instead, using Jean-Paul Sartre and Erving Goffman, I begin to discuss the existential account of agoraphobic self-identity in terms of the individual's relations to others. That is, to the performance of a specifically 'social' space drawing once more on the agoraphobic narratives of my respondents.

Notes

[1] Søren Kierkegaard's *Fear and Trembling: Dialectical Lyric* was first published in 1843 under the pseudonym Johanes de Silentio.

[2] '*That woman is more sensuous than man* appears at once in her physical structure' (Kierkegaard 1980: 65, original emphasis). The male is 'more qualified as spirit' (1980: 65).

[3] Clum and Knowles (1991) report that typically 59 per cent of panic disorder cases are female to 41 per cent male. See also Chambless and Mason (1986).

[4] The example Kierkegaard gives is of wanting to be, or be greater than, Caesar (1989: 49).

[5] '[A] fire takes hold in something that cannot burn, or cannot be burned up – the self' (1989: 49).

[6] This notion has currency in the field of market analysis, as evidenced by Solomon's *Consumer Behaviour*: 'A consumer exhibits attachment to an object to the extent that it is used by that person to maintain self-concept. Objects can act as a sort of *security blanket* by reinforcing our identities, especially in unfamiliar situations ... This coping process may protect the self from being diluted in a strange environment ... products are a sort of '*social crutch*' to be leaned upon during a period of uncertainty' (1994: 296/7, emphasis added). Solomon goes on to say 'A consumer's products place him or her in a social role which helps answer the question "who am I now?"' (1994: 296).

[7] Notwithstanding Gregson's criticism of the current narrow focus on the mall by theorists of consumption, for many agoraphobics this particular site, rather than 'the high street, speciality shopping complexes, discount warehouses, markets and mail order' (Gregson 1995: 136) does serve as the *epitome* of all others in terms of its ability to induce panic. It is for this reason that I take the mall as my focus.

[8] Mathews et al. (1981: 4) report from a study which found 'joining a line in a store' to be the most commonly reported anxiety-provoking situation amongst agoraphobics (96 per cent).

Chapter 4

'Putting on a Face': Sartre, Goffman and Agoraphobic Anxiety in Social Space

Hell is – other people! (Sartre, *No Exit*, Sc. 5)

Introduction

At the beginning of this book I argued that we must remain sensitive to the fact that individuals experience agoraphobia in different ways and in different social circumstances; that agoraphobia is a 'syndrome' best understood in terms of Wittgenstein's notion of 'family resemblance' (see introduction). This characterization is supported by the comments of those attending agoraphobic self-help groups who have spent many hours comparing and contrasting their experiences. As one agoraphobic argues, 'everybody gets their own, eh, agoraphobia [pause] it's a different creature to each different person that has it' (Carron). Another adds 'we're different people, it affects us differently, we all deal with it differently' (Moyra). Yet, as the last chapter illustrated, despite the intensely personalized nature of each experience and the apparent lack of a single causative factor, it is possible to begin to tease out certain threads that compose the matrix of this agoraphobic 'form of life'.

One of these, I argued, was the foregrounding of a kind of existential angst, a generalized concern about threats to the very nature of one's 'being', which become crystallized in the overwhelming and sudden onslaught of a panic attack. As Carron states, 'it absolutely petrifies you [pause] you think you're going to DIE'. The sufferer experiences feelings of derealization and depersonalization that seem to dissolve their identity. In Susan's words, 'the first [panic attack] was, I was about 18 ... I was on the bus ... and I got a' these weird feelings and I genuinely believed I was going to die'.

The bus where Susan experienced her panic attack is, like the shops and malls discussed in the last chapter, an example of the kind of intensely social space that seems to precipitate panic attacks. These social spaces thereby come to be regarded as threatening to the sufferer's feeling of 'ontological security', defined by Giddens (1997: 36) as the 'confidence or trust that the natural or social worlds are as they appear to be, including the basic existential parameters of the self and social identity'. Like many other fellow sufferers Susan, a recently married mother, responded to this threat by withdrawing into what she perceived as the sanctuary of her home in order to avoid the recurrence of such intense experiences. This contrast, between the anxieties associated with social spaces and the home as a place of relative safety, is another recurrent theme in agoraphobics' accounts:

> You're safe in your own home (Ann).

> When I was out [pause] I would take a panic attack, my safe place was my own four walls [pause] if you know what I mean, my house [pause] a safe place ... I'd be alright as long as I was in here (Maggie).

Often, agoraphobics speak of an urgency to get back home once a panic episode has begun outside, as Susan graphically illustrates:

> and here's another weird thing, you get nearer and nearer to your house, but your house appears to be moving further, and further, away from you. Like tunnel, sort of vision, and you get so freaked that you actually start physically running, I mean I've ran down that road before with my five year old practically scraping along the ground.

The fact that the home provides agoraphobics with a place of refuge from potentially corrosive contacts with others is readily understood if, as chapter 3 suggested, we characterize agoraphobics' experiences of panic attacks as indicative of a 'boundary crisis', as the problematizing of sufferers' ability to maintain and reconstruct their self-identity in the face of de(con)structive social influences. From this point of view the protective boundaries of the home exclude, or at least attenuate, some of the worst effects of social space. Maggie's 'own four walls' provide a relatively impermeable barrier that helps reinforce her psychocorporeal boundaries, a 'breathing space' within which she can recuperate. The home might be regarded as a kind of 'architectural' aid to re-constructing one's self-identity, an identity that is already more likely to be entangled within the fabric of the home for women than for men. In this sense the agoraphobic's avoidance of social space might be viewed as a retreat into a normatively feminine space where gendered boundaries are protected and potentially reinforced.

However, the agoraphobic's experience of home is often ambiguous; for some, home becomes so 'secure' that they are rendered incapable of leaving, they become, quite literally, house*bound*. Having lost their ability to negotiate their way in the world, and finding the locations in which they can move comfortably much reduced, sufferers may eventually be unable even to leave their homes without experiencing an incapacitating level of anxiety. The home is simultaneously experienced as both asylum and prison. As Maggie explains, 'I wouldn't go outside the door. I wouldn't even go to the back door to empty the bin'. Another respondent, Brenda, explains that if a neighbour passes the bottom of her garden:

> She would shout up to me cause she would know that I couldnae go down. You know how you see women blethering [chatting, gossiping] over the fences? Well I couldnae do that.

As we might expect, the actual intensity and form of such experiences vary. As Terry, one of the two male respondents points out, not everyone is so severely affected, 'everybody's no' the same, there's different degrees o' this [pause] agoraphobia'. For example, Ann explains, 'I was never housebound, but I did find the world was getting smaller'. Jane too states:

> I always loved going out, and even now, for me to spend a whole day in the house would be, torture. Even if it's just into the garden, I've got to get out for a wee while.

However, at the other end of the spectrum some find even their home provides little respite from anxiety and panic:

> When it was at its worst, I'd be at one end of the room, and my daughter would be in a pram at the other end, and I couldn't even cross the room to lift my baby out of the pram (Fiona).

Some agoraphobics, such as Iris, who for a time was confined to her bedroom, experience panic in all but one small part of their home and others, such as Linda, will panic even there.

There are a couple of points that need to be emphasized here. First, even if the individual does panic at home, it will often *still* be perceived as the safest place, because there, no-one else will *see* them panic (see below). Second, people clearly have differing abilities to cope with everyday aspects of social life and there is a continuum of experiences from the 'normal' (that is, 'appropriate' and manageable) to 'abnormal' (inappropriate and unmanageable). Thus while some respondents retreat still further into their homes, those more fortunate are able to extend the boundaries of the home outwards. For example, some agoraphobics regard their car as capable of providing 'ontological security' in much the same way as the home does. One sufferer does in fact describe the car as being 'like an extension of your home' (Nora).

This use of the home or car to reinforce the boundaries of self-identity might be seen in terms of Arnaud Levy's concept of 'subjective space': for Levy, subjective space 'surrounds us like an envelope, like a *second skin*' (Levy 1977, cited in Burgin, 1996: 213, emphasis added). The problem is, when the agoraphobic leaves home or gets out of her car, space ceases to be subjective. Once again she finds herself required to perform in front of those other people who create, determine and compose social space (Kirby 1996: 99). As the last chapter argued, sufferers are, it seems, too highly *sensitized* to the ways in which others 'charge' space through their presence. Social space is thus experienced as corrosive, or, we might say, *abject,* a term Julia Kristeva uses to denote that which we must 'radically exclude' from ourselves, but which is always present, threatening to invade us and causing *anxiety* (Kristeva 1982; Sibley 1995).

This chapter seeks to investigate further the nature of this problematic relation to social space and the difficulties agoraphobics find in its 'performance' (Freund 1998). In particular it questions why the routine activities that are hardly brought to consciousness in 'normal' experiences of social space become problematized as a source of *existential angst* for agoraphobics. There are thus two apparently contradictory aspects of the experience of social space that a theoretical account must address. On the one hand there is the question of the almost thoughtless 'practice of (normal) everyday life', and on the other the 'pathology' of an agoraphobic anxiety that, in certain social conditions, can erupt into a full-blown crisis of identity. For this reason, I shall draw upon theoretical perspectives that can speak to both the 'everyday' and the 'existential' aspects of social space, namely the sociological writings of Erving Goffman and the existentialist philosophy of Jean-Paul Sartre.

As anyone remotely familiar with the work of these figures will know, their perspectives are not straightforwardly compatible. However, both theorists do address what might be termed the social and spatial phenomenology of personal identity. While I do not intend to 'integrate' their theoretical perspectives, or play one off against the other, I will suggest that each has a different – though mutually beneficial – role to play in the elucidation of agoraphobic life-worlds. I thus aim to use complimentary aspects of their work to put forward an account that has both explanatory and (as chapter 7 shows more fully) therapeutic potential. By bringing sociological and philosophical perspectives regarding the problematic nature of social existence in contemporary western society to bear on agoraphobia, we might gain new and thoroughly *different* insights into the disorder that lie beyond the reach of, for example, clinical and indeed social psychology. (See, for example, Capps and Ochs (1995) for a contrasting, though insightful, psychological account of agoraphobia.)

Building upon my account of Kierkegaard's existentialism in the last chapter I first outline what I take to be the relevant aspects of Sartre's philosophy for understanding the existential backdrop to agoraphobic life-worlds. In particular I will draw out Sartre's account of the role of 'the gaze' of others in constituting (and threatening) self-identity. Sartre's abstract philosophical account is then contextualized and exemplified through the words of agoraphobic interviewees. I claim that their own words bear testimony to the existential anxieties underlying their experiences of social space and provide concrete examples of sufferers' fears of others' visual attention. I then attempt to connect Sartre's philosophical account to Goffman's empirically based work on individuals' use of strategies to protect themselves against the social intrusions of others. Again, drawing directly on agoraphobic narratives, I interpret their fearful and avoidant behaviour as a 'breakdown' in their ability to utilize the 'everyday' coping mechanisms highlighted by Goffman.

Finally I briefly address the potential *therapeutic* value of Sartre's and Goffman's insights. While Sartre demonstrates the need for awareness of and sensitivity to the profound and debilitating effects of social anxiety, Goffman discloses the practical means by which a degree of control can be reasserted over the subject's fragile social self. That is to say, by 'practising' the kind of coping mechanisms Goffman describes, the agoraphobic individual might gradually relearn to (at least partially) protect themselves from the perceived peril of the public eye. Taken together, then, the theories of Sartre and Goffman might thus help constitute a hermeneutics of agoraphobic experience, one that can provide practical insights into the problematic nature of social existence.

Sartre, the Look and Being-for-Others

The fear of fear, of the terror that erupts in the full-blown panic attack, rules the lives of severe agoraphobics. Such 'anxiety', 'anguish' or *Angst*, is (as we have seen) a unique and profoundly disturbing emotion yet it has, John Macquarrie writes, 'a subtle and elusive character that thought can scarcely grasp' (Macquarrie 1971: 128). Anxiety comes and goes. When present it seems to pervade everything and yet because of this its source is impossible to pinpoint. It's fear – of *what*?

> That which arouses anxiety is nothing, and it is nowhere. Yet ... it is so close as to be oppressive and stifling. It is not this or that particular thing, but rather the world, or being-in-the-world (Macquarrie 1971: 130).

For existential philosophers like Sartre (and Kierkegaard) nothingness is indeed the source of anxiety. It arises from the awareness that we exist but that we might cease to do so, that we might die. For Sartre, anxiety is 'the apprehension of nothingness' (1993: 29), the nothingness that 'lies coiled in the heart of being – like a worm' (1993: 56). For Heidegger too 'that *in the face of which* one has anxiety is being-in-the-world as such' (1988: 230, original emphasis). Anxiety both brings to the surface and problematizes the existence of the self and its world. It is an expression of the fragile nature of our individual existence, a mood that threatens 'something like a total disclosure of the human condition' (Macquarrie 1971: 130).

In a strange way '[a]nxiety individualizes Dasein'[1] says Heidegger. Anxiety is, he argues, the *Grundbefindlichkeit*, the 'ground' on which one finds or becomes aware of oneself as being, and yet paradoxically '*Angst* likewise is the experience of groundlessness, the absence of anything holding one in place and anchoring one's actions' (Cooper 1990: 130). Anxiety is that indefinite *edginess*, that vertigo inducing fear of the void that might at any time open beneath our feet.[2] With the onset of anxiety the familiar world recedes, its reality and concreteness are challenged and the subject is left feeling profoundly, inexplicably alone.

> In anxiety one feels "uncanny" [*unheimlich*]. Here the peculiar indefiniteness of that which Dasein finds itself alongside in anxiety, comes proximally to the expression "the nothing of nowhere". But here uncanniness also means "not-being-at-home" (Heidegger 1988: 189).

Although neither Sartre nor Heidegger mentions agoraphobia, this description of the subject's realization of the fragile and strange nature of their existence fits perfectly with agoraphobics' own accounts of their anxieties and (as we have just seen) in their case 'not-being-at-home' can be taken quite literally. In what she terms her progress diary Carron frequently writes of a world which appears (in a way akin to the clinical 'derealization' discussed in chapters 1 and 3) 'completely alien' and of an 'unreality' that can even extend to her own house.

So I'll go out, I'll take that bus journey. I'll go down and I'll come back. I'll walk in that door, and I'll be, it'll be like walking into a completely alien house, environment, the lot [pause] I'll know, that this is my house, that this is my suite, this is my, these, everything here is mine ... it was like this before I went out, but it'll – not – seem – *real*. And it's, it's like you're looking at things through water [pause] and it's the *scariest* thing in the world.[3]

The particular ab-normality of agoraphobic experience might be regarded as an extreme expression of the ontological anxieties that underlie all human existence, though the agoraphobic seems unusually exposed to and unprotected against the sudden onrush of anxiety. Yet such anguish is clearly absent from most people's everyday lives. This, Sartre contends, is because individuals are ordinarily shielded from 'true' awareness of the abyssal nature of our existence by their blind pursuit of a life immersed in the crowd. We pass our days in perpetual proximity to and in collusion with others, submerging our individuality and thereby refusing to recognize and take responsibility for our own ephemeral existence. This blissful ignorance is a 'false' way of life that, for Sartre, exemplifies 'inauthenticity' and 'bad faith'.

There is then a certain tension between the individual and the social nature of human life that can 'estrange' us from our potential as free, authentic, and independent individuals. 'My being-for-others is a fall through absolute emptiness toward objectivity ... this fall is an *alienation*' (Sartre 1993: 274-5). We can be 'taken hold of' by others, accept their views as our own, see ourselves as they do, and suppress our potential to be anything more. This, for existentialism, is life lived under the dictatorship of 'the they', as if our being-for-others, our *public* selves were all that mattered. We lose touch with our 'true' selves and become merely 'one of the crowd'.

For Sartre, existentialism presents a challenge to this homogenization of humanity, and provides the subject with the means to overcome their estrangement from their full human potential, to *resist* rather than *succumb* to the crowd. One of the ways in which we come to awareness of the inauthenticity of our being-for-others is precisely through the experience of *Angst*.

[I]n our everyday being-in-the-world and being-with-others we are able to tranquillize ourselves and escape the radicalness of the human condition. But anxiety jerks us out of these pseudo-securities. We are made to feel "uncanny" and "not at home" (Macquarrie 1971: 130).

Sartre's point is that anxiety is a necessary analogue of individuation:

What 'sinks away' in *Angst* is the world as interpreted by the "they". The usual meaning of things and actions fade as the everyday framework within which they have their slots becomes "uncanny" (Cooper 1990: 131).

Angst thus has the potential to bring one's estrangement from one's 'true self' to an end, shocking and propelling us out from a comfortable conventionality to face a new, more questioning mode of existence. For Sartre, this 'true self' from which we are 'alienated' does not already exist 'underneath', or hidden behind the false. Rather, it must be created, brought into being by *independent* action in the world, by increased awareness and realization of our potential for total freedom. To live authentically is an on-going struggle for freedom, a perpetual resistance to the pull of the crowd. 'Freedom that manifests itself through anguish, [angst] is characterized by a constantly renewed obligation to remake the Self that designates the free being' (Sartre 1993: 34-35).

Sartre, of course, wants the realization of 'authentic' existence to be philosophical rather than pathological, and agoraphobia should not be celebrated as a transcendence of 'bad faith'. The existentialist call to the individual is indeed to 'separate' herself from others, but while the agoraphobic subject may, in a sense, have taken *responsibility* for their social relations by trying to bring them to an end, this 'choice' does not stem from realization of personal *freedom*. In fact, an agoraphobic's avoidant behaviour might still be regarded as being ultimately determined by their relation to the crowd, though by their felt need to keep a distance from it. Thus, while such a life lacks the usual tranquillizing benefits that ensue from a sense of 'belonging' to (*being one of*), the crowd, Sartre might still ironically have judged such an existence 'inauthentic'.

The agoraphobic subject certainly feels far from comfortable in or cosseted by crowds. It is, as chapter 3 showed, social scenarios that usually provoke agoraphobic anxiety. But how can this discomfort be reconciled with Sartre's claims? The answer lies in the inherent ambiguity of anxiety and of human existence itself. There can be no Self without the Other but neither can there be a Self if one is wholly subsumed within the Other. Put differently, one must experience some degree of alienation *from* the crowd in order to overcome one's alienation from one's (authentic) self *in* the crowd. Ironically, it is our awareness of the 'objectifying' look of others that proves the source of our anxious striving towards self-recognition. 'I grasp the Other's look at the very centre of my act as the solidification and alienation of my own possibilities' (Sartre 1993: 263). Emotions surge through me as I am held by the look of others – anxiety, shame etc. – as I see myself as others must see me.

> I am ashamed of myself as *I appear* to the Other. By the mere appearance of the Other, I am put in the position of passing judgement on myself as an object, for it is as an object that I appear to the Other (Sartre 1993: 222).

These emotions are the indefinable and unstable 'ground' of my being, the source of my individuality.

> I now exist as *myself* for my unreflective consciousness. It is this irruption of the self which has been most often described: I see *myself* because *somebody* sees me – as it is usually expressed ... Behold now I am somebody! (Sartre 1993: 260/263.)

So Sartre's claims about the anaesthetic effects of the crowd have to be put in their proper context. Taking responsibility for oneself requires a difficult balancing act in our relation with others. 'Authentic being-with-others ... lets the human stand out as human in freedom and responsibility', while 'inauthentic being-with-others ... depersonalizes and dehumanizes [and] ... imposes uniformity' (Maquarrrie 1971: 91). Although the look brings us awareness of our separate existence, it also threatens to erode the very space within which we create ourselves. Without the look I am nothing, or at least not properly human, but with the look I become aware of the tenuous grounds of my own existence. The same look that solidifies and objectifies me also alienates, disturbs and threatens to dis*compose* me. (For explicitly feminist interpretations of Sartre and the 'look', see Murphy 1989, 1999.)

We are indebted to the Other for our sense of identity but, for this very reason, the presence of an-*Other* being has the capacity to destabilize our sense of ourselves and knock the ground out from under our feet. Sartre (1993: 255) explains that, when we are confronted with another person,

> suddenly an object has appeared which has stolen the world from me. Everything is in place; everything still exists for me; but everything is traversed by an invisible flight and fixed in the direction of a new object. The appearance of the Other in the world corresponds therefore to a fixed sliding of the whole universe, to a decentralization of the world which undermines the centralization which I am simultaneously effecting.

Their presence forces us to realize that there are 'other' ways in which the world can be constituted, 'other' configurations which we cannot control. By appearing before us, Danto explains, the Other initiates a 'struggle for centrality and ownership of worlds' (1975: 116). 'With the Other's look the "situation" escapes me ... *I am no longer master of the situation*' (Sartre 1993: 265, original emphasis).

This realization can be truly frightening, weakening one's sense of 'ontological security'. Some individuals seem never to be troubled by such intangible concerns, and it is clear from sufferers' accounts that there is no single factor that renders a particular individual sensitive and susceptible to the production of space by others. If Sartre is right, there should however be a clear relation between the individual's *self*-consciousness and their susceptibility to the *look* of Others. Further, as Sartre (1993: 29) infers, we might expect the (sensitized) subject to be disturbed by what she herself might (be seen to) do in the Other's presence. '[M]y being provokes anguish [*Angst*] to the extent that I distrust myself and my own reactions'. One might extrapolate from Sartre's assertion that '[v]ertigo is anguish to the extent that I am afraid not of falling over the precipice, but of throwing myself over', that agoraphobia is 'anguish' to the extent that I am afraid not of people themselves, but of what *I* might do in their presence.

Both of these predictions seem amply born out by interviewees' accounts. Without exception all those interviewed emphasized that it was the presence of other people that provoked anxiety. Thus Nora repudiates the common misconception of agoraphobia as fear of open spaces:

I'm not frightened when I'm *out*, I'm never happier than when I'm tramping about the countryside, but [pause] I think what they really mean is, if you're out with *people*, and maybe, you know, *trapped* by people.

It is important to emphasize that her anxiety does not stem from an inability to socialize but from a wariness of becoming the object of others' attentions. In common with many respondents Nora describes herself as 'very outgoing in some ways … if you speak to anybody who knows me, they'd say oh yes Nora she's great fun [laughs]'. She continues, however,

Inwardly [pause] its all an *act*, inwardly I'm really, I don't *ever* want to be the centre of attention … I'm the sort, if I had a heart attack in the middle of Princes Street [Edinburgh city-centre] I'd have to crawl into a back alley and die, rather than say to somebody could you get me help, because I would hate the attention.

Iris too expresses comparable concerns about the proximity of other people when I ask her to expand on her claim that 'shopping' is an especially difficult activity. She explains that, 'its em, the other people around. And the sort of noise, the noise levels, and lots of other people'.[4] While Iris' emphasis on the noise created by other people suggests that it may be necessary to expand Sartre's account of the gaze to account for other forms of 'sensory intrusion', the real point she is making here is one of loss of self-control in the crowd. This emphasis is reinforced later in the same interview: 'when I was home I felt I could *control* what was happening to me whereas if I went out, there were other people' – Iris breaks off here, and hesitates, before going on to tell the disturbing tale of having a panic attack in a crowded supermarket. Although her statement is left unfinished, it implies that, by their mere presence, others *rob* her of some degree of control over her world – that is, her self and her environs – in a manner reminiscent of Sartre's account.

Although some sufferers find it difficult to articulate exactly why other people are so disturbing, Brenda is unequivocal when asked if she can offer any explanation:

Aye. It's them looking at me. I'm scared if they look at me too much and guess how I'm feeling I'm gonnae burst into tears, cause I'd be totally embarrassed if that happened.

Brenda provides further insights when she describes how she became aware of the role of people *per se* in her agoraphobia. She states,

I was okay going out at 10 o'clock at night when it was dark and there was nobody there, take [the dog] a walk and it wouldnae bother me. That's when I realized it was people I was scared of as well.

Whereas being alone in a place traditionally thought of as 'public' when it is dark would itself be a source of anxiety for many women (da Costa Meyer 1996), Brenda's particular concerns for her personal safety are clearly of a different, one might say, *existential*, nature. It is the anxiety provoking *look* of others she wishes to avoid, and darkness can, in this respect, be seen to constitute a barrier to their gaze (Bankey 2001). Even if there are others around, Brenda will not be so readily exposed to their scrutiny; she is protected by the *cover* of darkness. Fran, an elderly StressWatch founder who became agoraphobic over fifty years ago, echoes this sentiment when she states that, while going out *anywhere* is difficult, 'it's easier to go to the pictures, because of the darkness, and you know no-one can see you'.

Brenda continues her account of the difficulties posed for her by a watchful public, describing her feelings about various swimming pools – clearly social places, but of a kind she *can* enter into under certain circumstances. She is relatively unconcerned by the attentions of those others in the pool itself, perhaps because she is almost entirely underwater and so largely hidden from their looks. What we find, however, is that her fears do incorporate the *possibility* of being subject to the Other's gaze (i.e. not one of the swimming 'crowd') from a distance. Brenda explains that it is thus important to find a pool where 'there's nobody staring in at you, and I don't need to worry about anybody looking in'. She illustrates by saying,

> I can go into [pool 'A'], I cannae go into [pool 'B'] just because it's a more open swimming pool. [Pool 'A's] closed in, you don't see so much outside, but [pool 'B's] got glass windows, you look through and people are sitting there looking, big glass windows looking into it, I cannae go in, I have tried.

Brenda's account repeatedly highlights her desire *not* to be subjected to another's look. An earlier comment revealed that this is at least partly due to a feeling of embarrassment, and a fearful preoccupation that she might do something embarrassing when the other is looking. This in fact is a concern she shares with other interviewees. Fran, for example, states: 'I got so as I wouldn't go out in case I did something embarrassing in company'. Jane, whose agoraphobia stemmed from an initial specific phobia of frost and snow, also alludes to such awkward concerns when asked why social spaces are so difficult:

> if it's a busy restaurant [pause] for instance going out for a Christmas dinner, and it is busy, beginning to feel, I want out of here, not wanting to embarrass yourself, or anybody else, and trying to figure a way of getting out of there [pause] its, it's the same with other social things, and I know when I'm approaching that kind of feeling, and I leave.

In response to a similar question about difficulties with social space, Nora says:

> I suppose that you feel that you're just going to completely lose control … and you're terrified you're going to fall over, or [pause] I suppose make a fool of yourself really … I suppose you just think that people will be watching.

Maintenance of composure can thus be seen to be significantly more difficult when the subject perceives that they are under the watchful eye of others, or as Brenda puts it (in a group interview) when 'all eyes are *on* you'.

There is then, a clear sense in which agoraphobic individuals are deeply disconcerted by the presence, and especially the *look*, of others. Drawing on Sartre's account, one might hypothesize that this anxious discomfort stems from a sense of objectification caused by that look. One simply cannot dictate how one's subjectivity is reconstituted in the other's eyes. A mere bodily air of civil indifference is either unmanageable or insufficient, and further strategies will be required, perhaps as simple in some cases as making the best of existing geographical features. Moyra says of *any* social space, but particularly in relation to her church (Moyra attends regularly, despite her difficulties – Christianity is clearly an important part of her life, and she stresses that her faith helps her cope with her disorder), 'I've got to sit at the back. I don't like thinking there're people behind me watching me fidgeting. I'm very self-conscious'. By sitting with her back to the wall, Moyra can minimize the potential for others to look at her, and thus also the discomfort that the thought of this generates. She is firmly guarded, from one direction at least, and need not be tormented by the potentially menacing content of the spatiality at her rear.

The other's look can seem to rob the individual of *vital* aspects of their identity, reducing their sense of embodied selfhood to that of an object over which they have only limited control. Sartre's account can help articulate the disturbing sense of alienation from one's *environment* that often accompanies the agoraphobic's experience of social space. On this view, as we have seen, the other effectively steals the world from the subject. Their simple presence alters the constitution of one's world, which no longer emanates only from one's self. These unsettling phenomena, of depersonalization and derealization respectively, can be among the most frightening aspects of agoraphobia for many sufferers. The perceived discrepancy between one's own subjective sense of reality at such times, and what one imagines to be the shared reality of others, can be deeply alarming, leading to horrendous feelings of abnormality and lack of control over self and world. Sufferers' interviews and diaries provide numerous examples of just how immediately this abnormality is felt. In addition to strange bodily sensations, distances are collapsed or expanded, and space becomes impossibly 'alien and strange'.

I want to turn now to explore Goffman's theorization of 'face-work' and 'front management', and question whether they can be understood as coping mechanisms for dealing with the spatially mediated social anxieties outlined by Sartre, and suffered to extremes by agoraphobics. I want to suggest that Goffman highlights these 'managing' behaviours, of which non-phobics are mostly unaware, but which, because they are both necessary and unattainable, loom large in many agoraphobic lives.

Goffman and Self-Management

If Sartre is right then there is a terrible weight of responsibility attached to our freedom; it is the freedom to take risks, make mistakes, to do wrong and to fail, to lose control and be judged for our actions. It is in this sense that Sartre believes we are 'condemned' to be free, doomed to make continual choices. Authentic human existence might then be seen as a perpetual process of 'crisis management'.

As is well known, much of Goffman's work constitutes a detailed study of interactions between strangers, in an attempt to uncover the means by which we *manage* ourselves in our relations with others. His analysis reveals a number of unspoken and largely taken for granted conventions, codes of conduct that govern our behaviour in the social realm and make it possible to remain comfortable in relatively close proximity with others. In Goffman's own words, '[w]hen in the presence of others, the individual is guided by a special set of rules ... called situational proprieties' (1963: 243). These conventions are intended to 'allow others their space' (Tseelon 1995: 38), to allow the illusion that one moves in a private, protective sphere, even when firmly entrenched in the social world of strangers. Conventions of social behaviour can then be seen to dictate respectfulness towards boundaries of personal space, a respectfulness of the right 'not to be stared at or examined', that must be reciprocal if social relations are to be unproblematic. This allows individuals to maintain some control or 'mastery' over their immediate environs, and limit the extent of alienation that the other's presence can inflict.

Perhaps foremost among those strategies Goffman's analysis highlights is that of 'civil inattention'. This entails

> giv[ing] to another enough visual notice to demonstrate that one appreciates that the other is present (and that one admits openly to having seen him), while at the next moment withdrawing one's attention from him so as to express that he does not constitute a target of special curiosity or design (1963: 84).

Civil inattention is the means by which we steer clear of the form of 'invasion' that is the direct gaze (Tseelon 1995: 67), and maintain the pretence of indifference to the appearance and concerns of others.

However, to be 'one of the crowd', to not stand out and thus be subject to intense scrutiny, one needs to behave appropriately, according to certain situational rules. 'Paradoxically', writes Goffman (1963: 35), 'the way in which he [sic] can give the least amount of information about himself ... is to fit in and act as persons of his kind are expected to act'. Of course, one needn't be aware of and intent on following explicit rules associated with particular situations. Rather, these often implicit behavioural standards and techniques of self-management are *internalized* so that we come to manage our behaviour, monitor movements and responses etc., in accordance with a tacit 'feel for the game', rather than by enacting formulaic codes.

The key to playing one's role in this 'drama' convincingly, and negotiating one's way around social space in a non-threatening and relatively inconspicuous manner, is a 'disciplined management of personal appearance or "personal front"' (1963: 24) and an ability to display 'sufficient harnessing of the self' (1963: 27). One's face must be composed into an 'inscrutable mask', indeed, we require a repertoire of faces, or masks, for a variety of situations, to maintain a comfortable degree of privacy. This principle of privacy might be defined as our ability to control the amount of information about ourselves that we display to others (Tseelon 1995: 74). 'Putting on a face' can be seen as one aspect of what Bauman (1993: 153) refers to as 'the art of mismeeting', a skill which *must* be mastered if one is to live among strangers. Bauman (1993: 155) explains the necessity of this strategy thus:

> The overall effect of deploying the art of mismeeting is "desocializing" the potentially social space around, or preventing the physical space in which one moves from turning into a social one.

Goffman claims, via his empirical observations, to have uncovered other specifiable tactics and devices that subjects employ to *limit* their involvement in social space and their exposure to its inherent dangers. The extent to which they use these 'involvement shields' (Goffman 1963: 40) and 'stalls' (Goffman 1971: 56) may depend, amongst other factors, on their level of social anxiety, but they are by no means merely the preserve of 'extreme' or 'abnormally' sensitive subjects. For many, such strategies are in fact a familiar part of everyday social life. Perhaps one of the most obvious everyday examples of the 'involvement shield' would be that of the newspaper used by commuters in a manner that precludes social exchanges with others. Similarly, 'stalls' protect and indeed *create* private space, by 'provid[ing] external, easily visible, defendable boundaries for a spatial claim' (1971: 57). The following example is provided: 'at beaches devices such as large towels and mats can be carried along with the claimant and unrolled when convenient, thus providing a portable stall' (1963: 56). In this manner, the individual can partially stabilize and control the 'ever shifting dimensions' of personal space (1963: 58).

Alternatively, subjects can territorialize space by attempting to highlight their need to *use* it for particular purposes. 'Use space', according to Goffman (1963: 58) is 'the territory immediately around or in front of an individual, his [sic] claim to which is respected because of apparent instrumental needs'. For example, giving 'sports [people] ... the amount of elbow room they require in order to manipulate their equipment'. Clearly, however, while some strategies may successfully assert legitimacy over a given spatial domain, they will not necessarily prevent the intrusive *look* of the Other. Goffman's example of 'use space' might in fact invite and encourage audience involvement.

It seems then that the social world is rife with danger, and coping with it requires *trust* that others are practicing civil inattention, that one is not constantly being stared at and invaded, and thus robbed of the protective barriers of personal space. What, though, would happen, were one to lose that trust in others, or could not trust one*self* to maintain a face or behave in a manner that does not invite unwelcome attention? Such a failure to manage social space has profound implications for the individual concerned.

> Failure or success at maintaining such norms has a very direct effect on the psychological integrity of the individual. At the same time, mere desire to abide by the norm – mere good will – is not enough, for in many cases the individual has no immediate control over his [sic] level of sustaining the norm. It is a question of the individual's condition, not his will; it is a question of conformance, not compliance (Goffman 1990: 152-3).

In terms of Goffman's (1969) dramaturgical analogy agoraphobics might be regarded as suffering from a kind of 'stage-fright', they shun 'public performances' where possible and strive to reduce their need to engage in 'impression management' to a minimum.[5]

Goffman's description of how one can be 'thrown out of step' with social norms does in fact capture something of agoraphobic accounts of panic, in particular, experiences of depersonalization. 'When an incident occurs and spontaneous involvement is threatened, then reality is threatened ... and the participants will feel unruled, unreal and anomic' (1967: 135). This threat of breakdown is, according to both Sartre and Goffman, more acute in the presence of others.

> The "strangeness" of strangers means precisely our feeling of being lost, of not knowing how to act and what to expect ... Avoidance of contact is the sole salvation, but even a complete avoidance, were it possible, would not save us from a degree of anxiety and uneasiness caused by a situation always pregnant with the danger of false steps and costly blunders (Bauman 1993: 149).

Each and every agoraphobic sufferer has learned from painful experience that '[t]he surest way for a person to prevent threats to his [sic] face is to avoid contacts in which these threats are likely to occur' (Goffman 1967: 15). However, complete avoidance of contacts is not always possible, for the '[d]efence of social space is never foolproof. Boundaries cannot be hermetically sealed. There is no really infallible cure against strangers, let alone against the dread they arouse' (Bauman 1993: 157). One might be required (for example by pain, or a strong sense of duty) to enter into social space for reasons as diverse as a trip to the dentist or, in the case of more than one respondent, a daughter's wedding. Failure to maintain one's 'composure' in these socially charged situations can lead to the individual having to struggle with what Goffman (1990) refers to as the 'stigma' of a 'spoiled identity'. Moyra explains:

> I tell as few people as possible, mainly because people don't understand it. And some people who I have told, nice enough people by the way [laughs] but you just dinnae see them afterwards, you know what I mean?

While the onset of agoraphobia typically entails a cessation, or at the very least, radical curtailment of the subject's involvement in social space, the necessity of a 'face saving' front will continue even within their newly constricted life-worlds. Contact with the 'outside world' may extend no further than relations with close friends and family members, but many make massive efforts to conceal their agoraphobia even here. Brenda says of her grandchildren, for example, 'I had to watch what I was doing so as I didnae frighten them as well ... I didnae want any of them to ever know how I was feeling'. One individual I spoke with was attempting to conceal the extent of her current difficulties in what one would imagine to be the reassuring environment of the self-help group, although for palpably different reasons. She says,

> I don't like to talk to them [other members of her group] about it being this bad again, so I go to them, *and I put on the face*, I put on this other, "oh yeah I'm fine", face. And I don't know why, I think its because ... I don't want to frighten anybody into thinking they're going to get this way again.

Once the process of putting on a face becomes a consciously felt need, then simply trying to fit in, to appear 'normal' in social space involves a substantial effort of will. Regardless of its purpose, maintaining a pretence of this kind is exhausting, and impossible to keep up indefinitely. A place to repose, and recompose, must be found, and though this often takes the shape of the subject's home, some may find temporary solace in other environments. Fiona, divorced from an unsupportive husband, and currently coping well enough to work part-time as a care assistant, explains that,

> even when I couldn't talk to anyone else, and when I'd be in the waiting room, keeping a smile on my face, as soon as I got into his [the G.P.'s] surgery I could really drop my guard, it was the one place where I could.

Many sufferers find that they need some kind of assistance to keep up their 'guard', and their (metaphorical and/or material) distance from others. A number of writers on agoraphobia have noted that sufferers tend to derive comfort from the use of dark glasses. Such items can be understood to function, in Goffman's terms, as portable 'involvement shields'. Though obviously less effective than a brick wall, glasses can provide an invaluable sensation of 'cover', both in the sense of protection, and of disguise. Commenting on their usefulness, Moyra explains that,

> it's partly cause it gives you a wee feeling of security. It's partly cause you think, well you know they can see you, but they cannae see your eyes. You know how sometimes if you're troubled it shows in your eyes?

The covering of one's eyes can partially obscure one's feelings, but it can also make it more difficult for others to intrude on, or 'trap' the subject, by attempting to engage her in conversation or some other form of exchange. As Goffman (1963: 95) points out, 'eye contact opens one up for face-engagement', and so dark glasses shield one from involvement by literally blocking the potential for eyes to 'meet'.

The pushing of an object such as a pram can also assist ease of movement through social space, an enabling tactic that is mentioned by several respondents. Moyra has said that 'I know I found it difficult when I stopped pushing the buggy, when my youngest son didnae need it anymore, I didnae know what to do with myself'. In Goffman's terms, we can understand this strategy by imagining that the buggy effectively creates an area of 'use space' around the subject. It signals to others the need to keep their distance, while at the same time conferring legitimacy on the subject's presence in social space. Perhaps the presence of children themselves can have something of a similar effect, blocking, or at least distracting attention from the anxious subject. Nora says of the time when her children were little, that 'really for many years, I thought that was my [pause] I hid behind them a bit'.

The need to hide, in one way or another, is powerfully illustrated by Brenda when she describes how she used the 'geography closest in' to this end. On her account, one's own body can be manipulated to constitute a form of corporeal defence against the other's look, though this is done with limited success.

> At one point I thought that's why I was, like I had let myself get quite big ... cause I was eating and eating and eating, and it was to stop folk fae staring at me. But then it was the opposite [laughs], like I felt then folk were staring at me *cause* I was so, big.

Brenda's bodily boundaries were thus felt to exceed expectations of normality, and to attract additional unwanted attention as a result. As Goffman emphasizes, the best defence against scrutiny is to not stand out.

The structure and production of space in sufferers' homes can also reveal much about the experience of this debilitating disorder. In Goffman's (1960) terms, the walls of the home demarcate (like the face) a 'front region' from a backstage haven from social pressures. Sufferers' homes are frequently organized to minimize the fear of the *look*. For example, Brenda's discourse on these issues followed a short (and barely articulate) question about the construction of safe space in and around her home – 'd'you feel as if you sort of do anything to build up sort of barriers or make it more like *your* home?'. In response, Brenda immediately embarks on an in-depth account of the socio-spatiality of her home-life:

> Right, see the sheds at the bottom of the garden, they were *right* out there [points to adjacent patio] so that people couldnae see me. That wasnae open like that, these sheds were up here.

In an effort to protect herself, Brenda arranged for two garden sheds to be positioned immediately outside her patio doors, thus completely blocking the view (both ways) out over her garden, towards the back fence and beyond. 'It felt too open for me [so] ... that's where the sheds were sitting, so that I didnae need to see, and I didnae think folk were, staring in at me'. Brenda's condition has improved since this time, but although her fear of being looked at has abated sufficiently to allow relocation of the garden sheds, she still makes use of more portable involvement shields in order to step outside. 'I can go out there and sit on the patio, put the patio-set out, and sit, but I've got the brolly on top of me [hand gestures over head] so, I'm sort of sheltered by the brolly'.

Brenda uses another 'visual aid' to screen herself from council gardeners who store equipment in a garage intrusively close to her front window.

> That's why I got these blinds was so I could do that … cause I get, nervous. Well I don't know who the men are … it's just [pause] they're in my, bit [makes 'guarding' hand gestures], and they shouldnae be there [laughs].

The tenor of Brenda's description suggests that her displeasure relates to their close proximity in itself, and not simply with their propensity to look, thus connecting with my next point. Steps were also taken to alter the geography immediately outside the front door, which opens directly onto the street. Brenda explains,

> that wall wasnae always there, and that got built so I could sit outside, and still feel I was inside the house. [The wall runs parallel to the front of her house, at a distance of several feet, and stands less than that in height.] The sun at night time comes up there [and so her husband said] come on out and sit, I says I *cannae*, that's fine he says, we'll soon sort that out [laughs], and he got that wall built [so] … that's worked, I can go and sit, I can even go out there now and wash my windows, because I'm still in my own [pause], I'm still in my own bit, I'm no' in [pause] there's nobody in my space [pause] in my air or anything, basically there's nobody in my space.

Brenda almost caricatures her own voice here, and uses her hands to symbolize *grasping* or possessiveness, as if self-mocking of a 'greed' for space, which she clearly perceives as 'irrational'. Her account makes plain that her sense of spatiality is fragile, and needs to be asserted and defended more explicitly than would 'normally' be the case.

I would suggest that Brenda's experience is a peculiarly gendered one, given that, in many contemporary Western societies at least, women are frequently more likely than men to be subjected to an intrusive and objectifying stare. This gendered phenomenon has been the focus of much feminist research, and theorists have explored the visual objectification of women from many angles (Bordo 1993; Tseelon 1995). One might surmise that the greater propensity for women to be stared at requires that they be adept at protective front management, that women tend to have a greater need to protect themselves from the Look than do men. For half of the population at least, 'doing gender' most likely involves an (ultra) awareness of the power of the Other's *look*, and this may go some way towards an explanation of the predominance of agoraphobia, as a spatially *and* socially mediated disorder, among women. After all, women are, in Goffman's (1977: 329) words, 'somewhat vulnerable in a chronic way to being hassled' (see chapter 6).

Conclusion

Although Sartre and Goffman approach the question of subjectivity in social space from different perspectives, both recognize its precarious nature. According to Sartre, the objectifying look of others is a defining moment in our becoming self-conscious; it is in their eyes that we see ourselves as somehow different, discontinuous from the crowd. But this discontinuity threatens to open a gulf beneath our feet; we face the abyss, the unfathomable anxiety of our own fragile existence. One response can be the bad faith of burying our heads in the sand, or more accurately, busying ourselves in the crowd. But even this supposes a certain immunity from the gaze, an ability to engage in what Goffman terms impression management, to 'put on a face' for others.

If instead, for one of innumerable reasons, we become hypersensitive to a gaze that only continues to heighten our 'self-consciousness', then social space threatens to become corrosive rather than constitutive of our identities. In this case, where the coping mechanisms of 'front management', 'stalls' and 'involvement shields' become incapable of suppressing the anxieties of everyday existence, then the agoraphobic takes refuge in the asylum of their home. And, like any asylum, as I have previously stated, its walls may come to demarcate both a place of safety and a prison.

Sartre celebrates the revelatory possibilities of an anxiety through which one finds oneself. However, Goffman adds a necessary corrective to this existential emphasis on authenticity by uncovering the workings of those more or less taken-for-granted mechanisms by which the individual maintains a sense of identity in the face of, and in spite of, the other's look. Were social spaces to become constant battlegrounds between competing compositions then even the strongest will could not sustain itself for long. There is a need to deflect, to *reflect* the look, and to protect oneself with the shield of civil indifference. When this shield fails, the subject experiences their very *self* as unmasked and naked, *exposed* to social space to a degree that makes their continued involvement intolerable.

It is in this context, once we have discovered just how and why social existence can be so un*nerving*, that sociological observation could be invaluable for the agoraphobic. Goffman's account, in particular, discloses those ordinarily taken-for-granted protective tactics that, I would suggest, sufferers must re-learn to enable (relatively) safe social engagement. This chapter has thus shown that, taken together, the aspects of Sartre's and Goffman's work that I have highlighted contribute to a hermeneutics of agoraphobic experience. It presents an account that is both 'abstract', in its recognition of existential anxieties, yet very much 'down to earth', identifying as it does the practical minutiae of psychological survival in the contemporary social world. I will return to the therapeutic potential of these findings in the final chapter and conclusion of this book.

I now want to turn from Sartre towards the related but more phenomenologically inclined and *embodied* philosophy of Merleau-Ponty. As indicated in the introduction, Merleau-Ponty expands Sartre's account of the gaze into a more general sensory phenomenology of the embodied individual, one that also recognizes the importance of the social (as opposed to merely individual) construction of lived space. In this sense chapter 5 continues the movement from an existentialism based on philosophical abstraction (Kierkegaard) towards an account grounded in everyday life and simultaneously helps prepare the way for the therapeutic considerations of later chapters.

Notes

[1] Dasein, literally 'there-being', is Heidegger's term for the human existant.

[2] Vertigo has provided a favourite metaphor in existential accounts of *Angst*.

[3] That Carron experiences discomfort even here, though clearly as a response to her venture outside, suggests that she clearly finds the home a place of only relative safety.

[4] The presence of others is clearly the predominant factor for this sufferer, but what Iris goes on to say lends support to the findings of this book concerning the disturbing and unheimlich architecture of consumer spaces. 'I think that, like being enclosed [pause] it's not so bad in a street, but you almost feel as if you're in a bubble, it's a totally unreal environment [pause]. It's alien, and strange'.

[5] Goffman (1969: 170) defines 'stage confidence' as 'the capacity to withstand the dangers and opportunities of appearing before large audiences without becoming abashed, embarrassed, self-conscious, or panicky'.

Chapter 5

A Phenomenology of Fear: Merleau-Ponty and Agoraphobic Life-Worlds

Introduction

I hope by now that I have demonstrated the relevance of existentialism as a framework for thinking about agoraphobia. Kierkegaard's and Sartre's works allow us to develop an interpretation of the phenomenology of the initial panic attack with its associated experiences of dizziness, derealization and depersonalization and offer profound insights into the intimate connection between subjectivity and social space that can help us understand its fearful potentialities. As the last chapter argued, these existential perspectives can be combined with other narratives, like those of Goffman, which also seek to elucidate the relationship between self and other in terms of the (usually unconscious) maintenance and control of boundaries.

However, as the previous chapters have intimated, there are still certain issues with Kierkegaard's and Sartre's thought that need to be addressed in order to ground their theoretical perspectives fully in everyday life. First, both philosophers often seem to regard the mind as something different from and somehow unrestricted by the fact of its embodiment. Despite their differences in outlook, both Kierkegaard (the Christian) and Sartre (the atheist) stress the mind's ability to transcend its current circumstances, whatever they might be, and make free choices. But while we may indeed be 'condemned to be free' in an abstract sense, the experiences of agoraphobics seem to challenge the idea that a refusal to face up to this freedom is 'all in the mind'. One of the most obvious aspects of sufferers' accounts is the manner in which agoraphobia manifests itself 'physically', in the body's suddenly increasing heartbeat, in sweating, palpitations, or legs that turn to jelly. Second, in his discussion of the influence of others, Sartre concentrates on 'the look' almost to the exclusion of other sensory contacts. Again, this seems consistent with a derogation of the import of bodily contact and embodied subjectivity.[1]

This chapter seeks to address the issue of embodiment by excavating and criticizing the Cartesian origins of this separation and privileging of the mental over the physical and by re-evaluating the role of other, non-visual sensations, in contributing to our performance of social space. It does this by drawing on the (existential) philosophy of Merleau-Ponty, which explicitly seeks to elucidate the interrelations between embodied selves and others in terms of a phenomenology of social space. I first set out my interpretation of Merleau-Ponty's approach, before attempting to demonstrate its particular power and relevance in relation to a case study, that is, a detailed account of the particular narrative of one individual agoraphobic, Linda.

As previous chapters have argued, there is no essential agoraphobic experience so choosing to emphasize Linda's account should not be taken as an attempt to 'reify' the narrative of one 'typical' subject over and above those of others. Rather, as chapter 2 suggested, the study of agoraphobia can benefit from the application of different methods at different times and this processual approach is mirrored in the form and content of the book itself. Since the purpose of this chapter is to draw together and ground existential accounts in everyday life, I have chosen to focus in detail on one respondent's account of their agoraphobic experiences. This allows me to present a more rounded account of agoraphobia, one which can integrate the various stages of the syndrome's development, from panic attack, to anxious incarceration in the home, to the mechanisms employed for coping with a seemingly hostile world in the narrative of a particular embodied individual.

While Linda's account certainly differs in detail from those of other respondents, the manner in which her own agoraphobia developed and impinges on her life adds a singular depth to the multiple voices heard so far. The danger with drawing on many accounts is that, in using these voices to illustrate and 'flesh out' philosophical points, one can sometimes seem to be abstracting them from their particular life-contexts; they too can thus come to seem almost disembodied. There is thus both an ethical and a methodological rationale for concretizing at least one voice within the context of a particular life history. In this way the reader is offered both an anchor and a point of contrast, a *different* approach that opens up, rather than closes down, the analytic and conceptual potential of the book as a whole.

The narrative of any of my agoraphobic respondents could have been represented here to fulfill this chapter's aims. That I chose Linda perhaps comes down to no more than her evidently (as we shall see) artistically inclined perceptual articulacy. This is especially relevant and informative in a chapter designed to highlight in phenomenological terms the mediating role of the sensations in acquiring and maintaining a sense of identity. Linda's discursive attention to her perception of colour and sound seemed to render her narrative extremely suitable for illustrating and concentrating the conceptual power of Merleau-Ponty's approach to understanding our place in the world. It is hoped that (this phenomenological reading of) Linda's story will thereby provide some contextualizing material with which fragments of others' stories and theories might be read and threaded together.

Merleau-Ponty, Space and the Horizons of Identity

> Every sensation carries within it the germ of a dream of depersonalization ... this activity takes place on the periphery of my being (Merleau-Ponty 1962: 215. Hereafter referenced as *PP*).

Since, as Merleau-Ponty points out, our own births and deaths are 'prepersonal horizons' that cannot be experienced by us, all that remains of our existence is our apprehension of ourselves as 'already born' and 'still alive' (*PP*: 216). But, although there are *defining* moments at our beginning and at our ending, the nature of our existence in between these points is less easily circumscribed. Modernity and modern philosophy, in keeping with Newton's solid mechanics and the possessive individualism of an emergent capitalism, have sought to impose hard and fast lines around us; to make of the individual an atom, a being-for-itself, autonomous, self-interested and introspective (see Irigaray 1985 and MacPherson 1979). For Descartes, the first modern philosopher, the *cogito* provided the acid test of our continued existence – 'I am thinking, therefore I am [*still in existence*]'. In this way existence was intellectualized, equated with the activity of thinking, of bringing our being before the court of self-consciousness.

Ironically, for a materialist age, where tangibility is regarded as the *sine qua non* of the external world's existence, the Cartesian solution redefines the self in terms of a life of the mind. Subjects are thinking things (*res cogito*) who are *not of this world*, souls without material substance (*res extensa*). In this way our identity is protected only at the cost of completely separating self and world, internal and external, the subjective from the objective. We are ghosts in the body's machine; disconnected observers watching the world pass us by.

But Merleau-Ponty argues that this cannot be. We are not a pure pre-given consciousness since consciousness is first and foremost 'intentional', it is always consciousness of some-*thing*.[2] What is more, this intentionality is not intellectually discovered but is something we become aware of through our *lived* experiences. Contra Descartes, consciousness is not something set apart from the world, but is itself a part of the world, it is 'a project of the world, meant for a world which it neither embraces nor possesses, but toward which it is perpetually directed' (*PP*: xvii). This involvement of the thinking subject in a project is paramount. For Merleau-Ponty, consciousness is 'in the first place not a matter of "I think" but of "I can"' (*PP*: 137). This dissolution of the absolute Cartesian boundary between self and world has a number of important implications. First, it re-emphasizes the body as integral to any understanding of the human situation. 'I am conscious of the world through the medium of my body ... I am conscious of my body via the world' (*PP*: 82). The 'subject that I am, when taken concretely, is inseparable from this body and this world' (*PP*: 40). We are *incarnate*: our access to the world is by way of our bodies. Second, this obviously means that, unlike *res cogito*, we do have extension, we *take up/occupy space*. Indeed, the experience of space is fundamental to our individuality, since our identity is no longer conceived of as a pre-given essence, a soul, but as a product and projection of our unique *situation* within the world. Third, thinking can no longer inoculate us against the world's intrusion since identity and thought are relational rather than absolute. 'I am not, therefore, in Hegel's phrase, "a hole in being" but a hollow, a fold, which has been made and can be unmade' (*PP*: 215).

Philosophy, for Merleau-Ponty, thus becomes an investigation into the fragility of our existence, into the complex inter-relations between body, space and self, into what it *feels* like, and what it means to be alive. Where Descartes sought understanding through the application of a philosophical method, an intellectual programme for pure thought, Merleau-Ponty directs his attention to the *phenomenology* of existence. It is not thinking but *perception* that is paradigmatic of our lived but never fully conscious experience. Rather than taking self (perceiver) or world (perceived) as givens, we must begin with perceptions themselves, that is, with phenomena. Perceptions are fundamental intentional states-of being, 'the background from which all acts stand out' (*PP*: x/xi) and which the ability to act presupposes. This 'bracketing off', or suspension of judgement about the ontology of self and world so as to focus on 'experiences', is referred to as the 'phenomenological reduction' or *epoché*.

While Descartes sought to introspectively establish the existence of an inner self through a thorough going skepticism about sensations, Merleau-Ponty regards these self-same sensations as the only access we have to our selves and our world. This is because sensations seem to occupy a liminal position. On the one hand sensations are the means by which I tentatively 'grasp, on the fringe of my own personal life and acts, a life of given consciousness' (*PP*: 216). They are the very ground from which my self-consciousness emerges to posit itself over and against the world. But, on the other hand, sensations remain ultimately intangible, I can never experience them as *mine* since they seemingly come and go without my calling them. Sensation 'runs through me without my being the cause of it' (*PP*: 216). I never *possess* sensations and in some strange way they seem to concern 'another self which has already sided with the world' (*PP*: 216). In a way sensations too are a kind of pre-personal horizon. 'I cannot know my own birth and death. Each sensation, being strictly speaking, the first, the last and only one of its kind, is a birth and death' (*PP*: 216).

Given the modern presupposition of an identity based upon the Cartesian self I 'naturally' assume ownership of *my* conscious thoughts since they seem to be constitutive of *my* identity. But if this identity is actually predicated on the ambiguous im-personal status of sensations then it is to sensations we must look if we are to understand who and what we are. 'Truth does not inhabit only "the inner man" [sic], or more accurately there is no inner man, man is in the world, and only in the world does he know himself' (*PP*: xi). Things are perceived before they can be thought and such perceptions evidence the complex intertwining of self and world. Phenomenology thus explores the essentially *perceptual* nature of our being-in-the-world, countering the post-Cartesian abstraction of mind from matter by concentrating on our most basic experiences, those sensations which seem to span the gap between internal and external. According to Merleau-Ponty all the efforts of phenomenology 'are concentrated upon re-achieving a direct and primitive contact with the world, and endowing that contact with a philosophical status' (*PP*: vii).

It is for these reasons that Merleau-Ponty is concerned to 'offer an account of space, time and the world as we "live" them' (*PP*: vii) rather than as abstract entities. Indeed, he introduces a crucial distinction between 'objective' and 'lived' space. The former is 'the space of rulers and tape measures', also described as physical, clear, geometrical or Euclidean space. The latter is the projection of our own spatial orientation on to the world, also referred to as subjective, existential, anthropological or phenomenal space.[3]

For Merleau-Ponty, the structure of *lived* space emerges from a relationship to our environs that is both perceived and produced through the activities we engage in. Lived space is 'cut up and patterned in terms of my projects' (Bannan 1967: 37). In terms of our earlier account, lived space is a performance. Our movement

> superimposes upon physical space a potential or human space ... The normal function which makes abstract movement possible is one of "projection" whereby the subject of movement keeps in front of him an area of free space in which what does not naturally exist may take on a semblance of existing (*PP*: 111).

Unlike geometrical space, lived space cannot be measured in meters or miles. Its distances are not fixed, but change according to our modes of existence, our moods, and environmental factors beyond our control e.g. whether it is night or day.[4] 'Our reading of the world' depends on the 'season of the year, and the season of being' (Evernden 1985: 131). Our sense of being-in-the the-world entails a kind of qualitative 'measurement' of lived space, a 'lived distance', which 'binds me to the things which count and exist for me, and links them to each other. This distance measures the scope of my life at every moment' (*PP*: 286).

As one would expect, Merleau-Ponty argues that lived space is mediated through the body.

> The body is the vehicle of being in the world, and having a body is, for a living creature, to be intervolved in a definite environment, to identify oneself with certain projects and be continually committed to them (*PP*: 82).

To act we need a kind of bodily sense of where we are, that is, a proprioceptive awareness of our physical deployment in and through space that usually operates below the level of consciousness. This unconscious awareness of our lived space comes through perception's 'operative intentionality', that *intending toward the world*, which 'evaluates the potentialities of my whole environment, so that objects appear as graspable or out of reach, inviting or threatening, as obstacles or aids' (*PP*: 440). We perceive and *live* this world through our various senses – seeing, hearing, touching etc. – all of which have their own characteristic spatiality. Because '[e]ach sense implicates the entire body, is intrinsically intersensory' (Zaner 1964: 181), these individual sense perceptions do, ordinarily, correspond and cohere, such that 'there is a unification of these spaces in one lived spatiality' (*PP*: 260). This sense of unification – of the senses, of spatiality, of existence as a whole – can, for Merleau-Ponty, largely be attributed to the workings of our intentional arc. He writes:

> [T]he life of consciousness ... is subtended by an 'intentional arc' which projects round about us our past, our future, our human setting, our physical, ideological and moral situation, or rather which results in our being situated in all these respects. It is this intentional arc which brings about the unity of the senses, of intelligence, of sensibility and motility. And it is this which "goes limp" in illness (*PP*: 136).

Lived Space and Perceptual Problems

Lived space is then distinct, though related, to the pre-existing and independent space of the 'objective' world. In Merleau-Ponty's own words, '[c]lear space, that impartial space in which all objects are equally important and enjoy the same right to existence, is not only surrounded, but also thoroughly permeated by another spatiality' (*PP*: 286). While objective space is 'always already there', we are involved and connected with it only by means of *lived* space, and this connection must be managed and maintained, 'ceaselessly composed by our own way of projecting the world' (*PP*: 287).

The 'synthesis' of space by the subject is 'a task that always has to be performed afresh' (*PP*: 140), a (usually unconscious) project of mediation that can never be completed once and for all. *Only* by means of this 'composure' of space do we come to distinguish our*selves*. 'Space is the correlate to, but also negation of, subjectivity', and so our understanding and awareness of what we are, and are not, is necessarily spatially mediated. It is drawn out of our *perceptual involvement* with the 'phenomenal field', from the 'dialogue or dialectic between organism and its environment, in which each *patterns* the other' (Spurling 1977: 11). Lived space is formed by, but also actively forms, our sense of self in the world though a 'pre-thematic "patterning" of the world that is difficult to catch at work' (Spurling 1977: 18).

From a phenomenological perspective:

> We have the experience of an *I*, not in the sense of an absolute subjectivity, but as one indivisibly demolished and remade over the course of time. The unity of either the subject or the object is not a real unity, but a *presumptive* unity on the horizon of experience (*PP*: 219).

Since our being-in-the-world must be continually negotiated, the effective construction and reconstruction of lived space requires a certain amount of 'perceptual faith' on the part of the subject. Our coherent sense of ourselves is by no means continuous or guaranteed and if we lose perceptual faith, our 'habitual' sense of ourselves in our milieu can become confused and senseless. Our *praktagnosia* (*PP*: 140), that practical bodily (rather than symbolically mediated) 'understanding' through which our body has access to its world, can be damaged or even lost. In the event, we can no longer trust that self and space will maintain normal, predictable relations, and we no longer know what to expect from the world – we might say that we have 'lost our grip' on it.

In other words, lived space is ordinarily 'taken-for-granted', produced below or beyond the level of conscious awareness and reveals itself only as a result of dysfunction, when it is 'thrown into relief by morbid deviations from the normal' (*PP*: 286).[5] So long as lived space is successfully maintained, we have a sense of freedom to move as we please (within the limitations of our physicality[6]), an easy relation with an environment that invites and encourages our involvement. But, if our sense of space breaks down, we may find ourselves facing a repellant, unwieldy environment that restricts our freedom of movement and threatens our sense of well*being*.

This conception of being at home in space has clear affiliations with Heidegger's notion of 'dwelling'. To dwell 'is the basic character of Being in keeping with which mortals exist' (Heidegger 1993: 362). *Wohnen* (dwelling) 'means to reside or stay, to dwell at peace, to be content' (Krell 1993: 345). Dwelling might be summarized as a form of 'sustainable personal development', of growing within and cultivating a sustaining relationship with one's environs:

> [O]nly because mortals pervade, persist through, spaces by their very essence are they able to go through spaces. But in going through spaces we do not give up our standing in them. Rather, we always go through spaces in such a way that we already sustain them by staying constantly with near and remote things (Heidegger 1993: 359).

Heidegger too suggests that moods and mental states might affect our ability to negotiate space, e.g. that depression might engender a 'loss of rapport with things' (Heidegger 1993: 359). Merleau-Ponty goes much further, arguing that '[w]hat protects the sane man [sic] against delirium or hallucination, is not his critical powers, but *the structure of his space*' (*PP*: 291, emphasis added).

For this reason Merleau-Ponty suggests that the best way to investigate lived space is through applying 'a new mode of analysis – existential analysis' (*PP*: 136) to the spatialities of 'deviant' modes of existence such as that associated with schizophrenia. Explaining the logic of this approach, Hammond *et al* (1991: 181) write that, '[t]he pathological case operates as a heuristic device that shocks one into an awareness of what is taken for granted. It is a means of gaining distance from the familiar, so that one is better able to explicate it'.[7] Within this context, Merleau-Ponty presents the reader with descriptions of disturbing scenes from schizophrenic life-worlds (drawn from the work of Binswanger and Minkowski) which his conceptual analysis aims to render comprehensible.

> A schizophrenic patient, in the mountains, stops before a landscape. After a short time he feels a threat hanging over him ... Suddenly the landscape is snatched away from him by some alien force. It is as if a second sky, black and boundless, were penetrating the blue sky of evening. This new sky is empty, "subtle, invisible and terrifying" (*PP*: 287).

The schizophrenic experiences a suddenly unmanageable transition within their lived space. The world becomes *unheimlich* as they seem to lose touch with the space of the objective world, and to be left alone in a now threatening and baffling subjective space.

For the schizophrenic, the world can suddenly change its physiognomy (or its 'living significance') and appear to be menacing and sinister.

> In the street, a kind of murmur *completely envelops him*; similarly he feels deprived of his freedom as if there were always people present *round about him*; at the café there seems to be something nebulous *around him* and he feels to be trembling [sic]; and when the voices are particularly frequent and numerous, the atmosphere *round him* is saturated with a kind of fire, and this produces a sort of oppression inside the heart and lungs and something in the nature of a mist round about his head (Minkowski, quoted in *PP*: 286, original emphasis).

Here, the subject's own lived space *limits*, rather than 'makes space for' his existence. It encloses and threatens to smother him, all the while separating him off from involvement with the shared space of others. In such cases of psychosis, the subject appears to have lost their sense of connection with objective space, and feels (with)drawn into their own fragmented and frightening spatiality. The ability to synthesize a tolerable compound of lived and objective space has been lost. We might say he has lost the ability to maintain the necessary balance or 'communion' (Rabil 1967: 32) between lived and objective space that allows for a comfortable, unthinking rapport with the world. This communion is, of course, necessary because 'I never wholly live in varieties of human space, but am always ultimately rooted in a natural and non-human space' (*PP*: 293).[8]

Merleau-Ponty argues that the ability to construct a 'homely' lived space is vital for one's mental health. If we are to feel that we belong, and have a place in the world, then we must be able to compose a suitable space. If we cannot 'situate' ourselves comfortably in relation to other people or things, we feel, quite literally, 'misfits'. This feeling can vary according to circumstances that are never entirely within our control.

> Sometimes between myself and the events there is a certain amount of play (*Spielraum*), which ensures that my freedom is preserved while the events do not cease to concern me. Sometimes, on the other hand, the lived distance is both too small and too great; the majority of the events cease to count for me, while the nearest ones obsess me. They enshroud me like the night and rob me of my individuality and freedom. I can literally no longer breathe; I am possessed (*PP*: 286).

This spatial understanding of mental health seems to have wider application. Consider, for example, the following instance of 'neurotic' experience.

> I felt as if I didn't fit into the world ... When I saw the snow, I felt I couldn't cope. One day it wasn't there and the next it was. I saw it and it upset me and I went to pieces ... I felt I did not want to be alive because I was not related to anything. I just seemed totally

out of everything and I started to cry. I couldn't cope with the hurt and the pain. I felt I would never feel part of anything (Sims 1983:116).

To keep our feet on the ground we need to be at ease with the space around us, to be on familiar and perhaps even intimate terms with our environs without suffering anxious 'feelings and fears of loss of self' (Sims 1983: 114).[9]

The Experience of Agoraphobia: Linda's Story

Merleau-Ponty's explication of the phenomenology of spatial distortions suffered in psychotic disorders (e.g. schizophrenia) and 'extreme states of consciousness' (e.g. neurological disorders and mescaline consumption) seems ideally suited for examining a spatially mediated disorder like agoraphobia. Sufferers' accounts of spatial distortions associated with agoraphobic attacks (see below), would also seem to make them perfect exemplars of Merleau-Ponty's phenomenological problematic, thereby placing us in a better position to judge his theory's philosophical merits. Much more importantly, if Merleau-Ponty's understanding of the relations between space, sensations and self-identity helps conceptualize agoraphobics' experiences and explain their symptoms, then it may be of some use in suggesting therapeutic or coping strategies for sufferers.

The best way to facilitate this *mutual* dialogue is by letting sufferers speak for themselves. In what follows I therefore focus on one particular sufferer's (Linda's) narrative in some detail. I do this for the reasons specified above and also because presenting a single case study might be regarded as the theoretical equivalent of an in-depth 'existential analysis' of the kind favoured by Merleau-Ponty, an analysis that would be less meaningful at a greater level of generality. As a painter and a musician Linda is also, as I argued above, especially well placed to articulate her sensory perceptions of an agoraphobic life-world. Her experiential account will, I hope, help to paint a sophisticated phenomenological picture of agoraphobia that foregrounds the relevance of 'lived space' in understanding agoraphobic and non-agoraphobic experiences.

I met 'Linda' through my ethnographic involvement with WASP, the self-help group she attended at that time. Linda is thirty-nine years old, single, and lives in a Scottish village with her two school-age children and an assortment of cats. She paints full-time, primarily to earn a living, but she also derives a great deal of pleasure from her work. Linda says that she has recently managed to strike a balance between painting that she actually enjoys, and that people are willing to buy:

> For years and years I painted flowers, and things that I [pause] och, that were very pretty, but they just weren't me. But, I think I've maybe cracked it with the balance, and it is land, its just landscapes [laughs] and it is so ironic, cause I, I've got an exhibition, in this place, ... the theme of the exhibition is travelling, [laughs] ... the irony of it just makes me smile, you know really, that here's an agoraphobic woman, producing these paintings of, you know [pause] far distant places.

As a child, Linda's parents had moved home frequently within Scotland. She recalls often feeling uprooted, and unsettled in a new town. Linda begins her narrative by describing these aspects of her childhood, and the fact that she does so suggests that the lack of a secure and stable base in her early years continues to hold a place of importance in her personal history. She explains that when she was seventeen, she planned to leave an area where she had recently moved and had few friends to work abroad as a nanny. It was at this stage in her life when she experienced her first panic attack: 'I can remember it quite vividly, it was the night before I left home'. When asked if she could describe it, she says:

> I was lying on my bedroom floor with headphones on, really really loud music, and em it's strange, hard to describe … it was just utter terror, you know, you might [pause] you might experience if somebody had a knife at your throat, you know, that level of fear.

Linda later comes to identify this as the beginning of her agoraphobia, and says 'I think, ever since then I've always been pretty miserable actually'. In Linda's mind, the original experience of panic is linked with her anxieties about leaving home. Although 'desperate' to leave, she was unsure and anxious about what departure from this place of perceived safety might entail. This ambivalence was a source of confusion, perhaps rooted in a sense of not really *belonging* anywhere. While the 'homely', non-social location of Linda's first panic attack is in itself unusual (illustrating the difficulty of positing 'essential' aspects of agoraphobia), the subsequent development of Linda's agoraphobic avoidance followed the fairly 'typical' route of gradually increased avoidance of places where she might *be seen* to panic (Bankey 2001).

The job abroad didn't work out, and Linda soon came back to live in various locations in England, reluctant to return to her parents' home. She continued to experience a fairly high, but variable level of background anxiety, and to develop agoraphobic avoidance of certain, mostly crowded places.

> I always had to live out in the country, I couldn't live in town, so it's, I couldn't walk through [pause] I mean I lived in Exeter, for five, six years, and I never once walked right across, right through the main street.

Although she did consult a doctor about her difficulties with negotiating social space, 'it was just a case of pull yourself together', and years passed before she came to understand what was wrong. 'I never came up with the word agoraphobia in all those years, I never knew what it meant, I just knew that I couldn't go, I couldn't get too far away from home'. When she did eventually come across the term agoraphobia, it was through her own reading; 'I know so much about it now, as much as any shrink probably, but then [pause] I just didn't know what the hell was happening'.

Despite an inability to stray far from the security of her home, she maintained a pretence of 'normality' for some years. In the midst of much discomfort and confusion Linda managed to complete a course at art college, clearly a difficult experience. She tried not to admit that she was struggling, and often adopted strategies, invented excuses, to disguise her difficulties.

> It gradually, it kinna got worse, and I was pretty miserable at art college. I hated it ... I was like the 'awkward' student, because, you know if they went on trips for instance, to France, or London, to galleries, I always said I couldn't afford it.

At the end of this period, she says, 'I'd had enough of England, and I wanted to come home'. We might interpret this 'home-sickness' as a need for stability in her life, perhaps a desire to 'put down roots', and in a sense this is what she proceeded to do.

Having returned north to Scotland, Linda says (my emphasis):

> I think I was quite happy for a while, quite content, I had a job, and I met this chap, and eh, my life was much more *structured* I think ... you get, you know, really *self contained* when you have children and a partner, but then ... after seven years, we split up, and that's really, when I went *downhill* again, my *foundations* were shaken.

Describing the early stages of the period following her return to Scotland, Linda emphasizes the positive aspects of social space, how the presence of others can in fact be experienced as helpful and *consolidating*. However, things clearly took a sudden turn for the worse with the break up of her relationship. A traumatic event such as this can subject almost anyone to a period of vulnerability and insecurity. Linda herself was deeply distressed by the experience, and she uses a powerful and *spatial* metaphorics to relate this part of her story. Her usage of the concepts of structure, foundations, containment, and downwards motion relative to her environment, can be interpreted in relation to Merleau-Ponty's phenomenological framework of identity, as a description of increased stability, then sudden erosion of her 'intentional arc' in response to changes in her emotional landscape. She had suddenly become very vulnerable indeed, destitute and dispossessed of self worth.

Shortly after her marriage break up, Linda had a major panic attack when she was driving, again listening to very loud music, and 'just collapsed really. It was like a snowball effect, an accumulation of all those years of ... not addressing things that were happening in my head'. She said she had been doing too much, and thinking,

> keep going, I can do it, I can do it, but actually, you know, inside, you're *crumbling* [pause] and all of a sudden, your body, mind and body and soul just, just *exploding*, and saying STOP, cause you're gonnae die, or you, something [pause] you're gonnae completely, *crack*, if you don't, address this.

The terms I've highlighted here emphasize the confusion Linda experiences in the boundaries of her phenomenal body, an *intolerable* and unsustainable confusion of internal and external space. She clearly loses the ability to project a protective boundary around herself and, as a result, is both assaulted by external space, *crumbling* inwardly under its pressure, and unable to prevent internal space from *exploding* outwards. Both sensations indicate a lack of containment, the *crack* that permits dispersion of self into its surroundings. Her account highlights the fact that personal 'integration is always precarious, since what can be patterned or structured can also be broken up as new patterns come into existence' (Spurling 1977: 39). While Linda experienced this attack, and indeed her first severe episode of panic, in an ostensibly 'private', non-social location, one might speculate that the loud music rendered her surrounding space virtually, if not *actually*, social. By this I mean that, even though the nature and presence of the 'noise' was of her own choosing, her emotional 'response' to the music was *amplified* beyond her control. At her most sensitive and vulnerable, even a virtual social presence was perceived as intrusive and frightening. Noise is, in any case, often closely associated with most forms of sociality (emphasized by Iris in chapter 4) – gatherings of people are rarely silent – and other respondents in this project refer to the distressing volume of the situation when describing the excessive multi-sensory stimulation social space seems to entail. (A phobic participant in a self-help video also reports 'the noise of the traffic' as disturbing, and that it 'seems to get louder and louder' during a panic episode – see chapter 7.)

So at this stage, when things had come to a head, Linda received psychiatric help, in the form of counselling and medication. Although appreciative of the attention to her problems, and relieved, to a degree, to have them out in the open, the treatment does little to alter the agoraphobic lifestyle she has maintained more or less consistently since the age of seventeen. When asked about the form of treatment she would liked to have received at this stage, she says:

> To be honest, in some ways, if I'd been taken into a hospital, put in a bed, and cuddled by the nurses, I mean that would have been great actually [laughs] to have a month or something out of the turmoil that my life was in.

Feeling, as she does, phenomenally insecure, Linda craves the tactility of human contact, cuddles that would soothe her enervated senses, and affirm her diminished sense of self. Placed in a consolatory environment, it seems she could find the time and space to re-*cover*, to strengthen her phenomenal bodily boundaries, and restore her weakened intentional arc.

For Linda, there are good reasons for *accepting* the debilitating nature of her illness. To do so means easing up on the constant battle against it, 'making the best of a bad lot' rather than trying to fight something she feels can't be beaten.[10] In this way, it is possible to avert the pain of panic attacks by completely avoiding situations that would provoke them. 'But the trouble with that is your world shrinks, till it is just the one room or your bed or whatever'. Linda goes on to say:

> The thing about agoraphobia, or maybe people with similar problems, is that they just jog along. There's a lady in the village, here where I live, hasn't come out her house in twenty years.

One can then, in a sense, 'choose' just to live within one's own 'particular radius', but this inevitably results in an extraordinarily restricted life-style in a constricted life-world.

Linda shares the view expressed by other sufferers that it is essential to *practice* going outside your home regularly if you are to avoid becoming house-bound for extended periods of time. This is expressed in a statement also reminiscent of my earlier usage of the quotation from Evernden, about the phenomenal, physiognomic relevance of 'seasons of the year', and 'seasons of being':

> Another relevant thing is the season. You know the winter [pause] I don't get much practice [pause] of you know going out cause the weather's crap, or whatever, you don't go for day trips or, you know, sort of hibernate. But in the summer, you know, the spirits lift, what have you, things become a lot easier, I become happier.

At different times, Linda stresses the fact that it becomes more difficult to function outside your home, or even to *imagine* doing so, if you 'get out the habit'. Confidence is easily lost, and painful to regain, so 'practice is really important ... I feel that I don't do enough, to be honest, but then, at the same time, I'm wary of not pushing myself too much'. There's a delicate balance to be struck in order to keep the physiognomy of your life-world relatively constant and familiar, to keep your involvement with it on anything like an even keel. Adopting Merleau-Ponty's terms, operative intentionality has to be *exercised*, you have to practice projecting and patterning lived space, with its potential for freedom of movement, to prevent it becoming hopelessly and debilitatingly contracted.

Linda recognizes that what she does for a living may not be all that conducive to developing and maintaining the skills necessary for coping in difficult spaces, which tend to be those populated with other people.

> Working at home ... it's not that healthy really, it's very isolating. And I think if you're out in an office environment, or a shop, or, you know that sort of thing, you know you're mixing with people just by the very nature of the job, you're getting practice every day.

The view Linda expresses here corresponds with some sufferers' theories about the development and exacerbation of their agoraphobic symptoms. Women have said that being 'stuck at home', for example, during the late stages of pregnancy, looking after young children, or in one case, because of a back injury, can, in a sense, cause you to 'forget', or lose the habitual bodily sense (*praktagnosia*) of, what's involved in going out. In a sense, you forget how to project your own lived space, necessary for freedom of movement, over and against those of other people. In such social circumstances, one's own lived space may need to be asserted more strongly, and defended against the seemingly corrosive presence of the lived spaces of others. There is then much effort, skill, confidence and (perceptual) *faith* required to convert space into something like a dwelling or 'homely' environment.

However, despite these perceived disadvantages of extended periods behind closed doors, the home clearly is often experienced as a place of sanctuary, an asylum rather than a prison (but see Valentine 1998). As we have seen, it is understandably tempting just to stay there, and avoid risking one's stability and security in the attempt to construct and explore a non-home based life-world. Most agoraphobics know, to their cost, the dangers of over-exposure to space outside the home. The world can become very frightening very quickly, and display an overwhelming physiognomy of horror. You have to tread lightly, and warily, to avoid coming face to face with the terrible reality of full-blown panic, a breakdown in relations between lived and objective space that causes one's surroundings to become threatening and awful. The subject of panic feels that the space they occupy is somehow qualitatively different from that of other people around them, they do not *belong* there, and their only option is to run, usually homewards. Having 'lost the place' in this way, home often seems to be the only imaginable space in which to rebuild your defences, away from the prying eyes and *presence* of others.

When asked if she can explain why the home is the most comfortable place to be, Linda thinks carefully about her answer. What she eventually articulates so clearly is in fact fairly typical of sufferers' views.

> When I look around here, kids aside ... it's like, almost like an extension [pause] of *me* [pause] you know it's, it feels safe because [pause] it's, it's a shield, it's a, protective thing, you know, obviously there's walls, and a roof em, but I would feel just as safe in my garden.

Her statement reveals that it would be simplistic to assume that the protection is derived from the physical enclosure of the bricks and mortar alone. While sufferers do, to an extent, *incorporate* these walls into their sense of self, to strengthen their phenomenal bodily boundaries, there is a deeper issue at stake here, relating to the structure and spatiality of our very existence. Merleau-Ponty's existential analysis reveals that home is often the centre of our lived space. It constitutes the heart of many of our life-worlds and provides a relatively stable base for us to orient ourselves in relation to. Agoraphobics, who spend so much of their time anxiously floating free from their moorings, are especially needy of this stability. At home, the subject can attempt to flex her operative intentionality, imagining lived distances opening out before her and 'gearing' herself towards them. These spaces can be physiognomically perceived as anything from an abyss that threatens to engulf, to an expanse that invites participation. The home provides relatively stable foundations from which sufferers can make judgements about their ability to cope, and to assess the prospective reliability of their intentional arc.

However, even here, as Linda discovered, there are times when the foundations turn out to be less than solid, and this realization can be truly terrifying. To lose the protective carapace of your sense of lived space *here*, to be invaded by *unheimlich* anxiety in your own home, is devastating to your sense of self. As Linda explains, when you're in a state of permanent anxiety, chronic, its really quite disabling, there's *nowhere* [pause], whether it's your home [pause], this is the scariest thing that I discovered, when I was really ill, it was, my home wasn't enough, my bed, my pillow, my mother staying here for months ['she was my rock'], wasn't enough ... nothing is, and, that's when you just lose it completely, you, there's just *nothing* there at all, you know, this fear.

What Linda seems to be expressing here is at the very heart of existential anxiety; the fear that at the root of everything is *nothing*, and that there is no place where you are safe from the threat to your existence, the loss of self, that hovers at the bounds of being. Linda survived this experience, though the level of anxiety seems to have persisted beyond what most sufferers of agoraphobia would have to endure.[11]

For agoraphobics then, anxieties are generally associated with attempts to project themselves beyond the home, which is ordinarily the safe centre from which their lived space radiates outwards. Linda notes:

I think everybody's got their own, radius, perimeter [laughs], mine is, I can go to [nearby town], it's a bit iffy if I can walk along the main street, an' stuff like, a Saturday [when it is busier] that's really pushing it.

Bannan can be seen to echo Linda's point about the home's importance (thus suggesting this is not a uniquely agoraphobic experience) when he says that 'the centre of my space is usually my home or dwelling, in terms of which I orient myself in the lived world' (Bannan 1967: 37).

One of the tactics Linda uses to extend the perimeter of her life-world beyond her home is to project a comparable sense of security onto her *car*, which she says she uses 'too much. Like when I park my car in the town, em I have to, be near the car ... there's always got to be this place, this sanctuary, where I can go and feel safe'. Linda is able to leave the house, and the radius of her life-world, her safety zone, extends far enough to include many of the things that are important for a 'full life'. 'The only thing that is missing in my life is this, ability to, to go, to say I'll go up north, for a holiday, and you know, what the hell, go to London and see an exhibition'. When she takes herself 'out of her element' and into unpredictable social situations, she is in a sense laying herself open to intrusion from the spatial compositions of others.

Linda has a high degree of perceptual sensitivity, and while this is clearly a source of pleasure, it is also disabling, in that she cannot always keep stimuli at a safe distance from her*self*. She often feels *invaded* by sounds:

Maybe it's being a musician I don't know, I can't read music, everything that I learn is just by ear, I've got one of these really intense, sort of listening, skills that I've got, it's a bit mind blowing at times ... if I'm in somewhere like Princes Street, where it's just all, the noise, the buzz, the general [pause] I think that definitely affects my brain, my mind, and it makes you very uneasy.

As a painter, Linda is also highly attuned, and sensitive to, visual stimuli, but she says, 'I wish I wasn't'. She feels unable to protect herself from their influence, or even assault; 'All my flipping senses are too, acute, for their own good ... everything about me's just really acute'. Linda is decidedly not 'thick-skinned', she emphasizes the close relationship between her *acute* sensitivity to sights and sounds and her inability to project a protective spatiality, or boundary. Uncalled for sensations seem to 'flood in' despite all she might do to try to shield herself from unwelcome stimuli. This is especially so in intense (social) spaces where these sensations emanate from and are produced by others yet seem to enter her very being in a manner completely beyond her control.[12] Here the boundaries of identity become confused and such sensations might even seem to belong to 'another self which has already sided with the world' (*PP*: 216).

Conclusion

Bryan Turner (1987: 218) has claimed that phenomenology has limited potential for the study of health and illness, as it can *describe*, but never explain the condition in question. However, Linda's account has shown that Merleau-Ponty's existential analysis provides much more than mere description. His focus on the continual construction of, and variable relations between, lived and objective space paves the way for a more *sensitive* reading of 'disordered' encounters between individuals and their life-worlds.

Merleau-Ponty offers a philosophical problematic that provides both a framework for understanding and a nuanced vocabulary capable of expressing the existential trauma and spatially mediated experiences of the agoraphobic. He also draws our attention to the vital point that those 'boundary' problems experienced by sufferers are neither so unique, nor so bizarre, if we choose to understand them in terms of a pathology of everyday life and existence. We are not Cartesian subjects. There are no fixed and immutable boundaries between self and world. The self is a project that needs to be constantly renegotiated and the senses provide the medium of this dialectic and of our orientation in social space. The experience of panic can arrive 'out of the blue' precisely because '[e]very sensation carries within it the germ of a dream of depersonalization ... [that] takes place on the periphery of my being' (*PP*: 215).

What is more, Merleau-Ponty also reminds us that we are not 'ghosts in a machine' but embodied entities. This suggests that any therapy that concerns itself merely with the mind (*res cogito*) fails to give due weight to the bodily and sensory aspects of the condition, to our 'extended' selves. What Linda wanted most was to be 'cuddled', to have her boundaries recognized and reinforced. Many other sufferers too find that they can gain a degree of security by hanging onto someone or something, as we saw in the previous chapter.

Merleau-Ponty's emphasis on the *sociality* of space also explains why we are all subject to the effusion of others' spatial compositions and how agoraphobics' characteristic and extreme discomfort in certain kinds of places – those commonly described as 'public' – can be indirectly derived from other people. (As outlined in chapter 1, I choose to describe characteristically 'difficult' space as 'social' rather than 'public', given the latter's problematic association with, yet distinction from, the 'private' sphere.) The *presence* of others and their projections can, for the agoraphobic at least, throw the stability of spatial situation into question. She can lose the essential sense that her bodily location is 'not one among many but the center in relation to every location' (Bannan 1967: 74). People 'perform' space, literally *populate* it with their own presence and constructions, causing it to become charged with their noisy, odiferous, colourful and tactile spatiality. The sensory phenomena that others make manifest has to be 'handled' by those who share their surroundings, and the 'performance' of this task is ordinarily taken for granted. The sufferer from agoraphobia, however, lacks the ability to assert their own subjective spatiality in the face of the spaces of others. When subjected to high degrees of sensory stimulation, sufferers feel 'trapped'. The contesting lived spaces of others jar their senses, and they can become overwhelmed and anxious to the extent that they suffer a full-blown panic attack. This chapter has shown that Merleau-Ponty's phenomenological framework is capable of expressing and assisting such frightening aspects of agoraphobic existence.

Merleau-Ponty's philosophy has been described as

> an attitude of wonder in the face of the world, a constant questioning and desire for understanding ... in the hope of attaining some kind of directedness or orientation. The goal of [his] philosophy is to enable each of us to rediscover his [sic] situation in the world (Spurling 1977: 5).

As I hope to have demonstrated, Merleau-Ponty can be read as initiating at least a partial 'explanation' of agoraphobia which builds upon the existential traditions represented by Kierkegaard and Sartre, and his quest for orientation may have an important part to play in wider geographies of mental health.

In the following chapter, I continue the exploration of embodied experience of social space and the subject's relations to others in terms of the rather extreme boundary changes experienced during pregnancy. What, after all, could challenge one's subjective boundaries more than having another individual develop within the limits of one's own body? Pregnancy also problematizes women's relations to social space in other ways, for example, in terms of their decreased mobility, a radical alteration of their body shape that often seems to make them a focus of public attention, and in terms of the altered responses of surrounding people. It goes without saying that pregnancy is a gender issue. However, the heavily gendered nature of agoraphobia, together with the fact that all of the women in my study groups had been pregnant (and that some of them actually associated their initial agoraphobic episodes with issues surrounding pregnancy) makes pregnancy of obvious import. Chapter 6 thus seeks to develop the discussion of Merleau-Ponty further by advancing a feminist interpretation of the possible synergies between the gendered experiences of pregnancy and agoraphobia.

Notes

[1] This is certainly not to say that Sartre ignores the issue of our embodiment (he does not) but simply to point out that his interests are primarily psychological and that for him our mental existence always precedes our essence. 'Man is nothing else but that which he makes of himself' (Sartre 1957: 28). The kind of (mental) freedom Sartre is interested in extolling is therefore somewhat different from the freedom desired by the housebound agoraphobic.

[2] Intentionality, from the Latin *intentio*, was a medieval term reintroduced by Franz Bretano in the nineteenth century who argued that all states of mind were directed towards, or about some-thing. It does not necessarily imply conscious intent on the part of the thinker (see Searle 1988: 3).

[3] 'Lived space' appears first, and most frequently, in his *Phenomenology of Perception*. Merleau-Ponty was not in fact the first theorist to use the concept of 'lived space', it was introduced by the psychologist Minkowski, in 1933, and incorporated some years later into the *Phenomenological Psychology* of Erwin Straus.

[4] Since 'spatiality is characteristic of our being in the world, there should be a variety of types of space according to the various modes of that commitment' (Bannan 1967: 99).

[5] In this sense, lived space can be conceptualized along similar lines to Leder's (1990) notion of the 'dys-appearing body', i.e. the body *re*-appears to conscious awareness as a result of some dys-function (see also Csordas 1994; Williams and Bendelow 1998a and 1998b, esp. ch.8).

[6] See Dyck 1995, Dorn 1998 and Park *et al* 1998 for discussion of 'disabled' experiences of space.

[7] Edie makes a similar point when he says that one should study extreme states of consciousness as 'it is sometimes easier to see in certain extreme cases what the normal or ordinary implies but hides' (Edie 1987: 104).

[8] It is vital to re-emphasize that phenomenology's emphasis on sensations is in no way indicative of idealism. Merleau-Ponty never contests the existence of a real material world. Indeed, he writes that 'phenomenology is a philosophy for which the world is always 'already there' before reflection begins – as an inalienable presence' (*PP*: vii).

[9] The experience of vertigo can also be understood in terms of a loss of *balance* between lived and objective space. (See Kirby 1996, Yardley 1994 and Quinodoz 1997 for accounts of 'vertiginous' experience of the world.)

[10] Many sufferers share the view that agoraphobia is something you never recover from, but merely have to learn to live with: 'I don't care what anybody says, you're never cured from agoraphobia' (Carron).

[11] Ordinarily, a panic attack will usually last somewhere between a few minutes and a couple of hours, though the intensity of the experience obviously makes it feel like much longer (Gournay 1989).

[12] One might speculate that the contrast between the depersonalization Linda occasionally experienced in *listening* to and being 'possessed' by recorded music and her ability to *play* music even in public places might be connected with the degree to which she regards herself as responsible for (and the source of) the sensations she experiences.

Chapter 6

Pregnant Pauses:
Agoraphobic Embodiment and the Limits
of (Im)pregnability

> The integrity of my body is undermined in pregnancy ... by the fact that the boundaries of my body are themselves in flux. In pregnancy I literally do not have a firm sense of where my body ends and the world begins (Young 1990: 163).

Introduction

In recent years, feminist geographer Robyn Longhurst has published a number of articles exploring the embodied experience of pregnant women, particularly in relation to 'public' space. (See Longhurst 1996; 1997; 1998; 1999; 2000a; 2000b.) By bringing the perspectives of feminist theorists such as Iris Marion Young and Elizabeth Grosz to bear on her own empirical research, Longhurst has demonstrated that women in contemporary western society experience an increased fluidity in the boundaries of their embodied selves during pregnancy. Further, she posits connections between these boundary phenomena and the increased discomfort and public censure experienced by pregnant women in the social sphere. One consequence of such 'politics of pregnability' that Longhurst explores is the occurrence of an, albeit temporary, 'shrinkage' in women's life-worlds.

In this chapter, I want to take up the project initiated by Longhurst in relation to some of the agoraphobic boundary issues raised in previous chapters. Agoraphobia is a disorder which, as I argue throughout this book, radically problematizes the subject's bodily boundaries. That is to say, it effectively disrupts the ordinarily stable, and largely taken-for-granted boundary between inside and outside, person and place. Certain environments are felt to be threatening and invasive, so much so that subjects increasingly withdraw from social space, describing their worlds as 'getting smaller' as their condition 'progresses'.

And so, according to Longhurst's account of pregnancy, and my own theorization of agoraphobia, both embodied 'conditions' apparently involve elements of boundary disruption and spatial restriction. Insofar as pregnancy also entails a retreat homewards in the manner outlined by Longhurst, away from the vagaries and demands of public life, it highlights some of those stereotypical aspects of femininity – domesticity, dependence etc. (Moi 1999) also expressed in agoraphobia. It is also interesting that, while the individual histories of those women who took part in my research on agoraphobia differed markedly from each other in numerous ways, they *all* had personal experience of pregnancy and motherhood that figured more or less prominently in their personal narratives of agoraphobia. I thus intend to open an investigation into the possible significance for sufferers from agoraphobia of these descriptive, and perhaps experiential, similarities, and will question whether the socio-spatial, psycho-corporeal experience of pregnancy and agoraphobia might have wider implications for feminist geographies.

However, I wish to emphasize at the outset that, although I tentatively suggest that there is some evidence of a synergistic relation between pregnancy and agoraphobia, I am in no way claiming that there is a straightforwardly *causal* connection between the two conditions. Rather, I seek to elucidate and interpret the meaning of commonalties in pregnant and agoraphobic experiences, through a phenomenology of women's relations to social space. The overwhelmingly gendered nature of agoraphobia suggests that there must be something about the cultural experience of *being a woman* in contemporary western society that contributes to its development. There must be aspects of 'doing gender' (Butler 1990; 1993), that for some women helps weave diverse webs of experience, yet binds them together in such a way that amounts to an agoraphobic syndrome. I aim to investigate pregnancy as one aspect – *but neither a necessary nor sufficient condition* – of an identifiably woman-centred phenomenology that might shed some light on the general complexion of women's experience that involves increased susceptibility to agoraphobia.

The (Im)pregnability of the Self

Previous chapters have described the distressing and boundary disruptive symptoms of agoraphobia in some detail, but for the purposes of this chapter, it is important to re-emphasize certain aspects of the phenomenology of the panic attack and sufferers' subsequent agoraphobic anxiety. Not only are the descriptions they provide indicative of an existential threat to the usually taken-for-granted boundedness of the self, but typically, the sufferer clearly expresses their experience in a language that freely mixes bodily and mental metaphors. In accordance with the phenomenological approach introduced in chapter 1, and detailed in chapter 5, the self *in question* is indubitably an embodied self. What is more, these experiences are elicited within the specific context of socially charged spaces, the agoraphobic 'comes out' in a cold sweat in the presence of, or in anticipation of the presence of, others.

If subjectivity and corporeality are inextricably intertwined then it seems obvious that changes in the latter will necessarily have some bearing on the former. This is, of course, the same insight that underlies attempts to theorize the experience of pregnancy. That is to say, the embodied experience of pregnancy will not leave the shape of a self unaltered; one cannot be the selfsame individual postpartum. Women themselves may be aware and wary of such change, as Young (1990: 168) in fact points out:

Especially if this is her first child she experiences the birth as a transition to a new self that she may both desire and fear. She fears a loss of identity, as though on the other side of the birth she herself became a transformed person, such that she would "never be the same again".

Pregnancy can, it seems, be a phenomenally intense and transformative experience, affecting, even 'refracting' one's identity in previously unimagined ways. In the words of Lucy Bailey (1999: 338/9),

[l]ike a beam of white light which is shone through a prism and emerges as a rainbow, a number of bright and colourful aspects of [respondents'] characters were arrayed as a result of the pregnancy.

Bailey reveals that, while some women felt ambivalent about their pregnancies, others obtained great satisfaction from discovering new aspects of their personalities, and described feelings of 'fulfilment' and increased self-worth. A number (albeit a minority) of the women taking part in Longhurst's research enjoyed the corporeal experience of pregnancy, and it was found to be liberating by some of the group interviewed by Rose Wiles (1994: 33/4). Wiles reported at the outset of her paper that '[t]he limited research literature indicates that pregnancy may be a time when women could experience their weight in a less negative way due to a greater social acceptability of fatness during pregnancy', though she concludes that this applies only to those she terms 'fat', not those of 'average-weight', who become pregnant. Wiles' research thus emphasizes that different women experience pregnancy very differently and ensuing changes are not necessarily positive.

Young's account of her own experience of spatial and bodily changes that take place during pregnancy are enlightening (for the uninitiated like myself at least), and lend some clues as to the emergence of embodied-identity issues for the pregnant subject. She states, for example; 'my belly swells into a pear ... this round, hard middle replacing the doughy belly with which I still identify' (1990: 162), revealing an emerging need to rethink one's habituated bodily image and attendant patterns of action. 'My automatic body habits become dislodged; the continuity between my customary body and my body at this moment is broken' (1990: 163). As a result of her changing body shape, the subject's previously taken-for-granted *sense* of her boundaries becomes increasingly unreliable. She undergoes a subtle metamorphosis that, though largely predictable, necessitates an increased awareness of and attention to her materiality and movement.

Gradually, Young explains, she becomes aware of the movements of the foetus, within, but somehow separate from her-self, an experience that unsettles her sense of individuality and coherence.

Pregnancy challenges the integration of my body experience by rendering fluid the boundary between what is within, myself, and what is outside, separate. I experience my insides as the space of another, yet my own body (1990: 163).

The pregnant woman undergoes what is surely a unique experience, to have an-*other* fully within her own body, 'inside and of me, yet becoming hourly and daily more separate' (Rich 1976: 47). She *is* no longer simply and unquestionably 'one'. How could such *inherent* ambiguity fail to perplex?

In addition to this sense of interior *difference*, Young describes significant alterations to her sense of being in place. She feels differently about her environment, and this is at least partly because she believes pregnancy entails that one's sense of the phenomenological location of the 'I' partially shifts from the region of the eyes to the region of the trunk. 'In this orientation that Straus calls 'pathic' we experience ourselves in greater sensory continuity with the surroundings ... Pregnancy roots me to the earth' (1990: 163). Clearly, this too can be characterized in terms of boundary alteration. Young feels less separate from what were previously her 'surrounds', and describes 'proprioceptive' disruptions, alterations to the sense of one's 'being in a body and orientated in space' (Csordas 1994: 5) that radically problematize one's conception and experience of being a *particular* situated self. Bailey (1999: 340) similarly found that her research subjects perceived a 'dissolution of the edges of their bodies'. She infers that '[f]or a pregnant woman, the edges of the self become blurred as the body no longer seems to operate as a physical marker of individuality'. To repeat Young's (1990: 163) compelling description, '[i]n pregnancy I literally do not have a firm sense of where my body ends and the world begins'.

So far, we have begun to get a sense of what pregnancy might entail for the identity of the subject in question, both in terms of direct personal experience, and in the more general, philosophically abstract sense of challenging traditional conceptions of selfhood. Longhurst is among those who argue that a non-dualistic understanding of pregnant embodiment undermines and *disrupts* restrictive (masculinist) notions of subjectivity. She writes (2000a: 55), '[p]regnant women undergo a bodily process that transgresses the boundary between inside and outside, self and other, one and two, subject and object'. In order to warrant these assertions, Longhurst presents an account of pregnancy that is less philosophically removed from everyday life than that presented so far. She, like Young, deals with bodily boundaries in some detail, but for Longhurst, these boundaries are a substantially more material and messy affair.

It is central to Longhurst's thesis regarding pregnant women's withdrawal from public space that the mere presence of their bodies – bodies that 'threaten to break their boundaries, to spill, to leak, to seep' – endangers 'rational public order'. Pregnant women, writes Longhurst;

> can be seen to occupy a borderline state as they disturb identity, system and order by not respecting borders, positions and rules. Their bodies are often considered to constantly threaten to expel matter from inside – to seep and leak – they may vomit (morning sickness), cry (pregnant women tend to be constructed as "overly" emotional ...), need to urinate more frequently, produce colustrum which may leak from their breasts, have a "show" appear, have their "waters break", and sweat with the effort of carrying the extra weight of their body (Longhurst 2000: 15).

This is an account of pregnancy that takes body boundaries and bodily fluids seriously. Drawing on Elizabeth Grosz, Mary Douglas and Julia Kristeva, Longhurst (2000a: 45) thus argues that pregnant bodies are discursively constructed as potentially 'leaky', inscribed as 'modes of seepage' and, as such, cannot be trusted to occupy public space. 'The pregnant woman who enters the public realm risks "soiling" herself and perhaps even others with matter produced by her body. Her body threatens to contaminate and to pollute". A number of Longhurst's research subjects thus describe their fear and disgust at the prospect of having to behave in such socially inappropriate ways when out in public as, for example, being sick in the gutter, or having their waters break in a shopping centre, events that would be 'revolting' or 'embarrassing'. Pregnant women do not want to be seen as 'out of control' in the presence of others, and their anxieties in this respect are reminiscent of those of agoraphobic women, who feel their boundaries cannot be relied upon to keep them under control and separate from the outside world. Indeed, pregnancy does seem to entail an analogous 'shrinkage' of the subject's life world as they withdraw from public places, and experience a partial 'confinement' in the private realm of their homes.

While it is certainly the case that pregnant women are fearful of what they themselves might (be seen to) do in public, there is also an awareness and apprehension of others' curiosity, their possible thoughts and behaviour in response to the subject's 'expectant' state. Pregnant women describe feeling subjected to the censorious and proprietorial behaviour of others, responses apparently most notable in social spaces such as bars, where subjects feel *judged* to be behaving in a morally inappropriate manner just by being there (presumably in the presence of alcohol and cigarette smoke; Longhurst 1999). This sense of public interest in, even 'policing' of, pregnant women's behaviour, can however be found to a lesser extent throughout social space, and is, arguably, simply an extension of the type of censure directed at women in general in a masculinist civic sphere. For many contemporary women, varying degrees of intervention and intrusion are an unwelcome reality of their day-to-day lives.

As I highlighted in chapter 4, it is the frequent fate of many women in contemporary western society to be subjected to and objectified by an intrusive male gaze (see Tseelon 1995). Young (1990: 166) stresses the positive aspects of the experience of the pregnant woman, arguing that '[t]he culture's separation of pregnancy and sexuality can liberate her from the sexually objectifying gaze that alienates and instrumentalizes her when in her nonpregnant state'. However, this tells only one side of the story. The *kind* of look directed at pregnant women is, admittedly, likely to be different, yet it could arguably intensify individuals' sense of being objectified, alienated, and instrumentalized by the other's gaze. While the pregnant woman is not, perhaps, as likely to be considered a *sexual* object, she is still the *object* of a different kind of look, certainly not fully a 'subject' in her own right. One could argue that she is attributed merely *instrumental* value, as the 'vessel' for her unborn child, and is thus, as Kristeva (1980: 237) contends, 'simply the site of her proceedings'. Insofar as women are thus objectified, they may be more likely to feel alienated than liberated by (some) others' attitudes to their embodied selves. The nature and intensity of the gaze directed at pregnant women may prove to be a source of disturbing self-consciousness, as they find themselves subjected to intrusions more unnerving than ever before.

It appears then that pregnant women are somehow 'answerable' to public concerns about their appearance and behaviour. They are expected to conform to normative standards of 'decency' to an even greater extent than non-pregnant women, who are clearly not exempt from public moral condemnation of their looks and behaviour. The female subject's usual sense of accountability is intensified as she is *reduced* to her role as expectant mother. This places added limitations upon her (already socially restricted) behaviour, dress etc. There are apparently appropriate ways to look and behave during pregnancy. While this situation is certainly changing, with clothing retailers increasingly responding to the demands of contemporary working and otherwise active pregnant 'consumers', it is still the case that women who stray too far from certain (dated) feminine ideals may be subject to severe social criticism. Pregnant women know that they are likely to 'provoke' strong reaction should they decide, for example, to 'parade themselves' in public in a bikini. As Longhurst (2000b: 11) reports, a front page newspaper photograph of women doing just that was followed by the publication of a number of 'disgusted' letters to the editor. (In one choice excerpt, the letter writer asks if these women were aware that 'they make special frocks for pregnant women to wear in public'. So distressed was s/he by their oversight, and keen to put things right, that s/he '*drew* frocks on all of them'.)

The women in Longhurst's study 'often claimed that they were uncomfortable with the idea of people looking at them [and this] in part stemmed from the feeling that their bodies were less attractive during pregnancy' (2000b: 6). The sense of exposure experienced by pregnant women can be aggravated by (what Kim Chernin (1981) identified as) our cultural 'tyranny of slenderness' (but compare Wiles 1994). At a time when they are more likely to attract attention, it seems many women have to contend with their own and what they perceive to be others' increasingly negative feelings about their enlarged bodies. Longhurst relates a number of instances where pregnant women express self-consciousness about their breasts and stomachs, or try to hide them from view, wearing baggy clothing to avoid attracting attention. One might also interpret this behaviour as an attempt to 'normalize' the appearance of boundaries experienced not only as 'bulging' (and 'therefore' unattractive) but also ambiguous and untrustworthy in that, as outlined above, they threaten to break down.

During pregnancy, women can be subjected to more than just the *look* of others. One of Longhurst's interviewees states that

> sometimes I feel as though being pregnant automatically deprives me of my individual identity and personal space. People seem to have a fascination with pregnant women's stomachs and want to pat them … because I've got a bump it seems that I've become public property (1998: 28).

In a strikingly similar vein, Bailey reports that

[a] number of the women observed that their bodies had become "public property", to be commented upon, patted and prodded, sometimes by people they knew very little or even complete strangers. They described their bodies as being "invaded", both by these other people and by the baby inside them (1999: 340).

That social restrictions on 'handling' others are apparently lifted during pregnancy means that visibly pregnant women forfeit part of their right to privacy, to keep strangers at a respectful distance, and that their sense of independence diminishes at the hands of others, sometimes quite literally. Pregnant women's 'condition' ostensibly confers rights on 'the public' to take an active and open interest in their bodies, not only by looking, but also by commenting on and even touching, behaviour that would not, ordinarily, be socially sanctioned.

Pregnancy and Agoraphobia: Commonalties and Potential Synergies

Given my characterization of both agoraphobia and pregnancy as posing a fundamental challenge to one's sense of self-identity via the problematizing of the self's previously taken-for-granted boundaries, perhaps we might be justified in looking for other links between these conditions.

It certainly seems from the work of Longhurst and others that many pregnant women find their increased exposure to public attention and potential censure problematic. The fact that pregnant subjects seem suddenly to lose others' respect for their personal space also seems to be of more than passing relevance for agoraphobic women (for whom privacy is a matter of psychological survival). Being the focus of others' concern is a *critical* fear and source of concern for agoraphobics, who want always to blend in, never stand out for social scrutiny.

An aversion to public attention means that, like agoraphobic women, many pregnant women also seem to withdraw into the confinement and relative security of their homes. There are of course some simply pragmatic or largely physiological reasons behind this partial withdrawal from public space. Pregnant women may feel tired more often, and encounter more barriers to social space arising directly from the built environment. Especially in the later stages of pregnancy, insufficient elevators, public toilets etc. may hinder women's movements, and the home may simply be the easiest and most comfortable place to be. Friends and family are likely to support this course of (in)action, helping to minimize the need for activity outside the home, by, for example, taking care of shopping, and so on.

Longhurst suggests that being pregnant may *permit* women to withdraw from public space in a manner that they may even find empowering. Clearly, as with agoraphobic withdrawal, retreat to one's home need not be an entirely negative experience. Apparently, though, it is very easy to quickly get out of the habit of negotiating social space. Presenting oneself appropriately in public is an *activity* that, while often taken for granted by non-phobic individuals, does need to be practised (Butler 1990, 1993): it takes discipline and control to blend in, and to feel comfortable with others. Perhaps for this reason some find that their difficulties extend beyond pregnancy when the initial reason for withdrawal is, literally, removed. In some circumstances the period of confinement seems to reduce subjects' confidence in their ability to manage themselves in the public realm. As one agoraphobic interviewee, Brenda, explains, 'it wasnae until I went to do it [leave home] that I realized, oh what's happened?'.

Given this, it might seem tempting to link pregnancy more directly with agoraphobia. Is it really accidental that, despite their vastly different life histories, all of the agoraphobic women I interviewed had experienced pregnancy? Certainly other studies have found that pregnancy is frequently associated with heightened anxiety. Capps and Ochs (1995: 32) recount the case of Meg who now regards these anxieties as a precursor of her agoraphobia.

> At the time I had anxieties but I hadn't become agoraphobic yet. I had, you know, the *glimmerings* of what was to come, but it wasn't full blown yet ... When I think about pregnancy *now* I have a lot more anxieties. Pregnancy *itself* is a form of confinement. You're in it for the *long-haul* once you start. It's *irrevocable* (original emphasis).

The majority of my own respondents also associated pregnancy with difficulties of an agoraphobic nature and, like Meg, some made direct links between their experiences of pregnancy and the onset of agoraphobic symptoms. Perhaps the clearest case of this was Susan who said that 'pregnancies definitely have made a mega difference to my life anxiety wise, it's increased the anxiety'. Susan explicitly identifies her agoraphobic symptoms as starting 'after the birth of my first child'. Since this time, she has experienced intermittent episodes of severe agoraphobia, linked, by her, to pregnancy. For example:

> I was fine for a good couple of years, and then I had my son, and I took post-natal depression again, and it just brought everything back all over again, and then it was pretty much an uphill struggle for the next [pause] five years.

I have met with and interviewed Susan on several occasions, covering the period before, during, and after her latest (and easiest) pregnancy, and we discussed the effect of this experience on her agoraphobia at some length. Although leaving home for Susan is never an effortless or thoughtless endeavour, she explains that she is even less likely go out when pregnant. This is partly, it seems, due to increased concern about others' perceptions and judgements of her size and appearance – not normally a major source of concern for Susan, who usually appears comfortable in tight-fitting clothes. She refers to

> getting like the side of a house [laughs]. It's no' the most attractive [pause] ... I'm no' a person that enjoys being pregnant, I'm no' into the clothing that you wear when you're pregnant [laughs] you know just silly wee vain things like that.

The kind of flowing and 'feminine' clothes usually deemed suitable for maternity wear are 'not quite me' for Susan, who is used to presenting herself in a more hard edged fashion. Although there is a sense in which she dismisses this aspect of her heightened sense of self-consciousness as simple 'vanity', there is surely more at stake here. Susan demonstrates that awareness of others' readiness to judge her will at the very least make it more difficult, if not impossible, to leave home. She seems hesitant about being cast as 'mum-to-be', a role she finds as ill fitting as the associated outfits.

Susan also exemplifies the importance of practising the negotiation of social space, explaining that during her most recent pregnancy, she was in fact able to attend hospital, as opposed to the home visits she required previously. She suggests that her experience was so different because she felt much 'stronger' at the outset of this period of potential confinement, having made concerted efforts to 'exercise' her ability to leave home. As she explains; 'before I fell pregnant I'd done a lot of hard work', but even so, still 'had a few setbacks during the pregnancy'. Other recovering agoraphobic women frequently comment that it is essential to expose yourself to the outside world on a regular basis. To fail to do so for any reason, whether through fear of panic, physical impairment, or simply due to bad weather, can only have a detrimental effect on their confidence in their ability to handle themselves in public.

However, despite these testimonies, it would be extremely unwise to jump to the conclusion that pregnancy *causes* agoraphobia. For one thing, there are many other issues at stake surrounding the transition to being a parent and the experience of early motherhood in itself which ought to be taken into account. It is often not pregnancy itself but its aftermath that is cited by respondents as a crucial period in the onset of their agoraphobic symptoms. When I first interview Lizzy, for example, (an original member and regular attendee of WLMI) and ask how and when she became agoraphobic, she states:

> It was actually really, when I had my children [pause] that it came. I think it did … When I came out of hospital, and a' the anxieties of having your first child, I just, I just felt [pause] very anxious about him, thought he was going to stop breathing [pause] But nobody was there to reassure me, you were on your own, and that was it. And ever since then I was always very wary about taking him out on my own, travelling on the bus.

When I asked Fiona to elaborate on the increased agoraphobic difficulties she explicitly associated with pregnancy, it emerges that for her too it was rather the experience of early motherhood that posed the greatest problems. She states that, *on top of everything else*, having a

> baby to cope with just toppled me. When it was at its worst, I'd be at one end of the room, and my daughter would be in a pram at the other end, and I couldn't even cross the room to lift my baby out of the pram.

In the following exchange (which took place at a later date), Fiona raises issues that potentially complicate understandings of her near paralysing panic.

> Joyce – You know we talked about pregnancy before?
> Fiona – Yes, that's when it got bad, but I don't think it was the actual pregnancy …
> Joyce – Right. It just sort of happened around that time, you wouldn't say there was anything to do with, about being pregnant?

Fiona – No I don't. Although having said that, maybe the way I handled it was [pause] because your hormone levels and everything are all, you know [pause] But I think it was [pause] it wasn't pregnancy itself [pause] ... em, it got so bad at that time, it really got so bad when I had the baby, em, I mean I couldn't go out. You know you used to have those vans, shop vans that come round? I couldn't go out to that.

Fiona is clearly reluctant to 'blame' pregnancy for her difficulties. But, despite her seeming willingness to take personal responsibility for her agoraphobia, she strives to articulate her understanding of the relation between 'the way [she] *handled*' pregnancy, and hormonal fluctuations associated with the experience she presumably could not *help*. One can imagine that popular discourses constructing pregnant (and indeed *all*) women as at the mercy of raging hormones may complicate and undermine Fiona's, and other subjects' attempts to understand and manage their complex emotional worlds during any difficult time. Pregnant women, perhaps especially those who are already emotionally vulnerable, may be hindered in their attempts to 'handle' their situation by repeated suggestions that they are probably 'going nuts' (Longhurst 1997).

Fiona, who, like the majority of agoraphobic respondents, has had to contend with debilitating fears about 'losing it' or 'cracking up' – especially in the early days of her disorder – is by no means alone in attaching significance to hormonal fluctuations. Moyra, for example, states that

obviously with a female it depends on the time of the month as well, you know sometimes if you're having a bad spell? It can be that time of the month, and it can be weird if you're pregnant as well.

Mary, too, raises the issue of hormones, but then in seeming contradiction to Moyra's position, states,

I was fine when I was having at least *one* of my children ... [so] it can't all be hormonal. You get men that come along [to the group] as well. It can't just be the female hormone [laughs].

The discussion soon turns to pregnancy with Brenda, too, when, during an interview in her own home, her daughter Jane arrives on a visit with her own two-month-old baby. In the lengthy conversation and humorous exchanges that follow, Jane describes her experience of 'going do-lally' – a term she readily applies to her mother – during her own pregnancy. She in fact experienced a panic attack for the first time when she was pregnant, and could now, she claimed, understand from personal experience why her mother would 'run round like a headless chicken'. As an aside, intended to illustrate her daughter's unflattering metaphorical account of panic, Brenda describes how, during a 'nightmare' attempt at a shopping trip two years previously, she

knocked a pregnant woman flying in the middle, crossing a street in Glasgow, and I turned round to apologize and I done it again ... *that's* what she means by going round like a headless chicken. I just didnae know where I was to go, where I was gonnae feel safe.

As we can see, since her own pregnancy Jane now has some insight into her mother's confused and disjointed (over)reaction to 'unsafe' social situations. Perhaps her own (necessarily limited) background knowledge of what panic entails helped Jane to prepare, to manage her experience and prevent it getting 'out of hand'. Jane may now associate the strange spatial experience of panic with pregnancy, and thus with aspects of her *female* 'situation'. She knows, however, that she was not *really* 'going nuts'. Benefiting precisely from the experience of women who have gone before, Jane knew that life, and relative stability, goes on.

The idea that there might be a straightforward causal link between pregnancy and agoraphobia is further undermined by the very diversity of causal narratives offered by sufferers themselves. It is surely understandable that agoraphobia sufferers want to isolate the cause/s of their initial experience of panic and subsequent development of agoraphobia, and that they are willing to devote considerable time and energy to the attempt. However, despite their best efforts, few eventually find themselves in Susan's situation, able to specify a particular *reason* for their disorder. Rather, sufferers are likely to point to a number of potentially stressful factors that existed or occurred prior to the onset of their condition, and to speculate that their agoraphobic symptoms manifested themselves as a result of some combination of these factors. Each individual has a different story to tell, and many continue to speculate and re-work their narratives to the present day.

While pregnancy was a theme that frequently surfaced during discussions relating to the issue of possible causes, it was far from being the only one. Among those diverse events respondents put forward as causative or contributory factors of their own agoraphobia, one respondent, for example, refers to the sudden and distressing appearance in her neighbourhood of a man resembling the perpetrator of earlier abuse. Another (Carron) described the stress of being detained behind Checkpoint Charlie for several hours, in addition to maintaining an extremely demoralizing relationship with her mother-in-law. A third respondent (Fiona) struggles to identify any possible cause, and tentatively suggests that she may have suffered some underlying trauma from witnessing bombing raids as a child. These accounts are not 'typical' and, if anything, agoraphobic causal narratives are similar only insofar as they *don't know* exactly why they became agoraphobic, there tends to be something of a consensus that 'you can't put your finger on any one thing' (Fiona).

In their attempts to understand their situation more fully, and hoping to gain insights that might aid recovery from agoraphobia, many subjects have turned to a variety of clinical literature and self-help resources (discussed further in chapter 7), only to find that they shed little more light on the subject of causes. In line with respondents' accounts, such sources often acknowledge that stressful events – such as a family bereavement, period of illness, change of home/job, or pregnancy – may be considered a significant or precipitating factor (Marks 1987; Gournay 1989). Carron states, 'I've read a lot of books on it, an' they do say it's an emotional reaction to something, it's a reaction to something *really* emotional'. On the other hand, it is also acknowledged that panic may seem to have appeared entirely 'out of the blue'. If, though, we consider the nature of some of those stressful events highlighted more closely, we might speculate that they often involve *lack of control* over one's situation, and seem to call one's place in the world into question.

We can speculate similarly about the phenomenal impact of other significant life events, that they involve the subject in facing substantial change, to one's relations with people and places, and perhaps also to one's prior role. In the face of bereavement, for example, one feels utterly impotent. The profound experience of loss can seem to alter the shape of one's family relations and life-world, forcing the subject to negotiate new spaces, or gaps in their lives where previously there were none. Certain other events outlined by sufferers may seem to do so only temporarily, even 'trivially', as in the case of short term physical disability such as that caused by a back injury (suffered by Jane). However, even despite apparent recovery, this experience might have lasting impact on the way the individual conceives of herself, for example, in terms of her physical capabilities and independence. It may make her feel *powerless*, shake or undermine her self-confidence and ontological security, in a deep-rooted and enduring way.

What then are we to make of these accounts? One could argue that such situations, including pregnancy and maternity, insofar as they entail a radical rethink of one's life-world in potentially disorientating ways, could fall within Karl Jaspers' phenomenological conception of 'boundary situations'. Such boundary phenomena, for Jaspers, can be defined as 'situations I *cannot get out of*, situations I *cannot see through as a whole*' (Jaspers, quoted in Young-Bruehl 1981: 21), and which, as such, threaten one's 'orientation' in the world. Boundary situations challenge our 'space for freedom' (Young-Bruehl 1981: xi). They cause us to sit up and take note of our bearings, to question ourselves and (our place in) the world and perhaps wonder how to go on. According to Alfons Grieder (1999: 186), boundary situations are thought to arise, for example, 'through serious accident, conflict with others, illness, having to face one's death, or the death of somebody very close'. While pregnancy is often strongly desired, and thus not an 'adverse' event of the kind most often associated with Jaspers' conceptualization, experiential accounts reveal that it can indeed present a challenge to the subject's 'space for freedom'.

Such 'space' is, as feminist geographers recognize, and Young famously demonstrates in *Throwing Like a Girl* (1990), already far more restrictive for women than for men. I would thus argue that 'boundary situations', pregnancy included, *heighten* women's awareness of existent, intensely limiting aspects of a masculinist social world, bringing home (sometimes literally) the socio-spatial ramifications of women's culturally situated, gendered embodiment. A woman's identity, perhaps especially her identity as a pregnant woman, is already decided, or at least only minimally negotiable, and experiencing such limited choice about how to re-present oneself socially is bound to be unnerving, that is, to stimulate 'nerves' about 'public appearances'.

The commonalties between pregnancy and agoraphobia and the apparently synergistic relationship between them in certain cases might then be better understood as phenomena that are, quite literally, *engendered* by women's experiences of subjectivity in contemporary society. The social pressures experienced by women, whereby they are expected to be especially sensitive and responsive to the needs and gaze of others, especially in terms of their orientation toward their family, children etc. might all contribute to the difficulty of maintaining a stable sense of self. This is nowhere more the case than in pregnancy and childcare. It obviously takes a degree of *self*-confidence in relation to the social world to trust one's capability in relation to another being. In the absence of a solid and reassuring support network, the pregnant or newly parental subject can easily feel isolated and incompetent, and perceive their behaviour to be policed to a greater extent. Two of the respondents in this project, for example, described concerns about being, and being *judged* to be, 'unfit' mothers (one, Mary, worries that her children 'brought themselves up') and independently described periods of particularly intense anxiety in anticipation of the arrival of health visitors in their homes. In fact, as each subsequently became less and less likely to leave home – more house*bound* – they became increasingly house-proud, keeping their own domestic territory as near perfect as possible. These subjects describe real concerns about 'doing' mothering and housework badly, and failing to perform stereotypical aspects of their gender role to culturally expected and internalized (high) standards.

I would however like to close with the following, somewhat tentative, suggestion. If we can continue to develop and strengthen feminist understandings of embodied subjectivity that take women's 'other' socio-spatial experience seriously, we would perhaps be less likely, as a culture, to pathologize boundary difference and disruptions of the kind undergone by pregnant, 'insecure', and otherwise 'excluded' subjects. That is to say, a model of subjectivity that is not based on masculinist conceptions of bounded autonomy may allow (especially female) subjects to experience boundary changes as something other than *crises*. One could then feel *ambiguous* or uncertain about one's new and shifting place in the world, without feeling intolerably exposed or unstable.

Conclusion

This chapter initiated an inquiry into conceptual and socio-spatial links between the experience of two conditions overwhelmingly considered 'women's troubles', and aspects of the peculiarly gendered treatment of women in contemporary western society. Drawing on theoretical accounts and empirical research with women who are pregnant, women who are agoraphobic, and women who have experience of both conditions, the chapter suggested that each presents an intensification of and is synergistically linked with the meaning and doing of gender.

It was shown that agoraphobia, like pregnancy, demonstrably troubles the subject's sense of being a discrete individual, reliably separate from the 'outside' world. In both conditions, women experience their boundaries as more fluid, less protective, and untrustworthy; they fear they might 'break down'. Additionally, they may feel (and indeed, be encouraged to feel) less rational and emotionally stable. Both states also involve heightened awareness of the censorious and objectifying gaze of others that is associated with reluctance to present oneself in public. Moreover, the chapter has argued that pregnancy and agoraphobia serve to magnify pre-existent socio-spatial difficulties of womanhood, chiefly, that of maintaining a bounded and secure sense of identity, in addition to a 'space for freedom', in limited and limiting social circumstances.

Feminist theorists (e.g. Battersby 1998; Bordo 1995) have convincingly argued that women generally are subjected to ideological imperatives to look and act according to certain gendered norms, behavioural standards that still lean towards domesticity and servility as opposed to authority and sociality. They have further shown that women tend to be represented as having more fluid boundaries than men, and as such, they are less suited to, even threaten the stability of, the rational order of the masculine public realm. Women's 'volatile bodies' (Grosz 1993) thus require a degree of 'confinement' within the domestic sphere. In pregnancy and agoraphobia, this feminine boundary situation apparently reaches something of a crisis point, as the need for confinement intensifies. That is to say, in both conditions, women are treated to '*more* of the same', rather than differently, and it is, I argue, the socio-cultural situation of women in general that 'produces' agoraphobia.

At the outset of this chapter, I suggested that there must be something about the existential realities and meanings of *being a woman* that entails susceptibility to agoraphobia. I aimed to show that it is a 'condition' – perhaps (following Bordo on anorexia) an 'overdetermined symptom' – of the cultural situation of women. Bordo has written:

> I take the psychopathologies that develop within a culture, far from being anomalies or aberrations, to be characteristic expressions of that culture; to be, indeed, the crystallization of much that is wrong with it (1995: 141).

Aside from being a powerful argument for the need to study such psychopathologies – they hold 'keys to cultural self-diagnosis and self-scrutiny' – Bordo's basic contention is that disorders such as agoraphobia are potentially *understandable* in the light of wider social circumstances. Agoraphobia is a *cultural* condition, coalescent from the treatment and experience of women, and must be explored by, amongst other approaches, a phenomenology of femaleness and femininity. By focusing on one aspect of some women's experience, this chapter has gone some way towards extending feminist explorations of pregnability initiated by Young, Longhurst *et al*, to probe affinities with agoraphobia. It has thus begun to forge experiential and conceptual links between some aspects of 'doing' and experiencing gender (particularly those highlighted by pregnancy), and the gendered distribution of agoraphobia in contemporary western society.

In the following chapter, I want to explore the issue of treatment for and possible recovery from agoraphobia touched on in chapter 4. Specifically, chapter

7 will offer an experientially grounded critique of Cognitive Behavioural Therapy (CBT), a method of treatment commonly advocated by widely used and increasingly available forms of self-help resources. This final chapter is motivated by a desire to reveal the ways in which CBT apparently employs a peculiarly gendered and restrictive, rather than liberating, model of the self, one that relies on the kind of Cartesian dualisms that I have worked to undermine throughout this book. The chapter aims to look beyond such potentially debilitating views of selfhood, and gesture towards a more spatialized notion of subjectivity that could be appropriate, comfortable and comforting for the respondents re-presented by this study. Such a potentially enabling conception might, I suggest, help integrate the fractured spaces in, of and around agoraphobic selves with which this book has been centrally concerned.

Chapter 7

'All in the Mind...?':
Analysing the Subject of 'Self-Help'

I think they try to fob us off, I don't think, em, they really take us, very seriously. I think they think probably because it's mostly women that go to them with these problems that they think we're, a bit neurotic, maybe just a bit weak, an [pause] don't want to cope with, maybe everyday life (Kathy, agoraphobia sufferer and contributor to self-help video 'A').

Introduction

It has been clear from the outset of this book that it is 'peopled', rather than 'open' spaces that agoraphobics fear and avoid, though this clearly runs contrary to both popular opinion and dictionary definitions of the disorder. As we have seen, the lack of understanding surrounding agoraphobia can combine with sufferers' own reclusiveness to make their condition extraordinarily isolating, both physically and emotionally. Sufferers may be reluctant to discuss or even admit to having difficulties they themselves don't understand, even with close friends and family. Recall Moyra's poignant observation, quoted in chapter 4, that many 'nice enough people' in whom she had confided responded by keeping their distance – 'you just dinnae see them afterwards, you know what I mean?'. Should sufferers eventually decide to seek help from health-care professionals, it is, as Kathy makes plain in the opening quotation (above), by no means guaranteed that help or even understanding will be forthcoming. In circumstances such as these, self-help resources may seem to be the only source of information and support available, and it is therefore important to understand their character and role.

'Self-help' resources in general are, as Robin Allwood remarks (1996: 19), 'aimed primarily at women', being mostly concerned with what are (stereo)typically considered to be 'women's problems', those related to depression, weight and anxiety. (For feminist perspectives on therapy for women see Brown 1995, Brown and Root 1996 and McLeod 1994; on feminist psychology see Burman 1998 and Unger 1998.) Such texts would therefore seem to be an important focus for feminist researchers' attention, and I want to consider what a feminist interpretation of self-help resources aimed at agoraphobic women might reveal. Perhaps the most immediately obvious focus for such an analysis relates to the ambiguity inherent in the very name of the genre. The term 'self-help' can either be read in the ('common'-) sense of 'help yourself', or in the alternative, more complex sense of 'help for the self'. The difference here lies between what is straightforwardly 'self-service', and what is a kind of therapy for the treatment of selves. Both senses clearly harbour implicit definitions, but the latter in particular begs specific questions regarding the nature of assumptions underlying this definition. Namely, given that there are contested notions of what a self is or can be, whose idea of self is being invoked here? To exactly what kind of self is help is being offered?

Previous chapters have specifically criticized the hegemony of what I referred to as a Cartesian model of the self, one that separates mind (*res cogitans*) from body (*res extensa*) and consciousness from the world. I have argued that such a separation is both untenable and incapable of understanding the nature of the boundary crises that agoraphobics experience. A key part of this book has been to develop a different understanding of the self and of self-consciousness as embodied, intentional and as a 'performance' in social space. As chapter 1 pointed out, there is a consensus among contemporary feminist theorists that subjectivity and our sense of self is thoroughly embodied, that the body and not just the mind is 'a medium of culture' (Bordo 1993: 65). (See also Battersby 1998, Butler 1990; 1993, Irigaray 1985.) Further, as chapter 5 argued, the existential phenomenology of Merleau-Ponty and others entails a critique of the pure consciousness of Cartesian philosophy positing instead an 'intentional' consciousness that is not something set apart from the world, but is part of the world. Thus, in Merleau-Ponty's words, our conscious being is 'a project of the world, meant for a world which it neither embraces or possesses, but toward which it is perpetually directed' (*PP*: xvii). The agoraphobic's difficulty lies, I have argued, in the dangers inherent in maintaining the 'semi-permeable' boundaries of the self in the face of a (social) world that sometimes appears as if it threatens to possess their very being, to embrace them so tightly as to squeeze them out of existence.

We have then two quite distinct models of the self and yet, ironically, self-help materials currently available to agoraphobics tend to elide any attempt to bring the predominantly Cartesian model into question. As we shall see, they tend to discuss issues about self-identity and self-control in entirely physical or mental terms – as dependent upon *either* a 'strong constitution' or a 'weak will' rather than, as previous chapters have argued, an inextricably complex combination of both. The representation of agoraphobic selves on offer in self-help materials is set *against*, while dependent upon, a myth of 'normal' (Cartesian) identity.

The audience of the self-help videos I discuss below are shown to be subjected to the pervasive notion that there is a particular version of self, a form of 'normality', to which they should aspire. This mythical model is, I will argue, inextricably bound up with an interlocking set of masculinist discourses, and is overly, if not overtly, reliant on an individualist, yet dualistic, liberal humanist model of the self. Maintaining the hegemony of this Cartesian conceptualization involves the denigration and suppression of alternative discourses, those associated with the wrong (read feminine) side of a familiar set of hierarchic dichotomies; of the body (as opposed to mind); of nature (held against culture); and of emotion (as always subservient to reason). (See for example Bondi 1992, and Plumwood

1993.[1]) I argue that these dualistic discourses are instrumental in positioning the subject(s) of agoraphobia, and have an intrinsically *normative* role and rationale.

My ultimate aim here will be to show that this limited and limiting 'ideal' cannot contain or account for, and thus excludes, the embodied experiences of agoraphobic women. One need not, however, necessarily respond to this situation by positioning the agoraphobic as pathological, by representing her in terms of an aberration of an assumed normal self, and one that requires re-moulding within Cartesian terms. One might rather question the validity and adequacy of the Cartesian model itself. I would in fact suggest that sympathetic analysis of sufferers' relations with their environment might offer a challenge, a possibility for subverting a model incapable of recognizing this alternative (and largely gender specific) experience, elements of which, I contend, are common to non-phobic women as well.

One outcome of pursuing this line of thinking will be the suggestion that self-help resources drawing on dualistic discourses are of limited effectiveness for sufferers from agoraphobia. This conclusion begs the question of what self-help of a more inclusive hue (perhaps informed by feminist critiques of singular notions of self) might look like. I thus intend to close this book, but obviously *not* the debate, by gesturing toward possible alternatives.

Approaching the Subject of Self-help

The texts I will analyse in this chapter are two self-help *videos* of very different kinds.[2] *Not all in the Mind* is an 'amateur' production, collaboratively made by eight previous members of WASP (video A). It is twenty minutes long, and just over two minutes of that time is given over to 'expert' opinion. This is in marked contrast with the second video, which continually intersperses two sufferers' accounts with the opinions of two health-care professionals. *Fight or Flight* (video P) was professionally produced in Australia by 'Monkey See Productions', and lasts for forty-eight minutes. Before considering the videos in more detail, I will first outline the methods of data collection, interpretation and analysis that are used in, and particular to, this chapter: how does one attempt to investigate and analyse underlying presumptions about the nature of self in any given text(s)?

My approach to the analysis of these videos draws principally on the methodological model developed by critical discourse analyst Norman Fairclough.[3] He situates 'text' within the realm of 'discourse practice' (text production and consumption), which is in turn situated within the realm of 'sociocultural practice'. His depiction of the relations between these elements is intended to convey their interdependency. It suggests that we must look wider afield than the environment immediately surrounding the text if we are to appreciate the complexities of its role, and uncover its unacknowledged relations with wider socio-cultural and political factors.

Fairclough describes the *critical* aspect that distinguishes his approach as being concerned with 'not just describing discursive practices, but also showing how discourse is shaped by relations of power and ideologies, and the constructive effects discourse has upon social identities, social relations and systems of knowledge and belief' (1992: 12). More recently, Fairclough (1995) has tended to focus on the analysis of 'communicative events' such as newspaper editorials and television documentaries. With respect to their representations of the world, Fairclough states that

> media texts do not merely "mirror realities" as is sometimes naively assumed; they constitute versions of reality in ways which depend on the social positions and interests and objectives of those who produce them (1995: 103-4).

Various, but unacknowledged choices will have been made, and it is the task of the analyst to bring these choices to light, to reveal

> what is included and what is excluded, what is made explicit or left implicit, what is foregrounded and what is backgrounded, what is thematized and what is unthematized [and] what process types and categories are drawn upon to represent events (Fairclough 1995: 104, cited in Duley 1997: 42[4]).

Fairclough's contextual approach suits the purposes of this chapter. As I have indicated above, the excluded, implicit, backgrounded or unthematized elements I am concerned to bring to light in this analysis relate to notions of selfhood, and in particular, the ways in which selves are conceptualised in dualistic and gendered terms. Fairclough's method enables a socio-philosophical, or, I would suggest, a *geographical* reading that can unearth the unacknowledged discourses of self that underpin self-help resources.

The decision to focus on video recordings, rather than audio or printed literature, requires some explanation. Books and pamphlets especially are perhaps more commonly used than any other self-help resource, and they certainly tend to be more accessible in terms of both availability and cost. The main factor influencing my neglect of these alternatives is that video material is the most amenable to a shared and participatory 'reading'. We could watch, and comment, on these texts together. An additional reason for favouring the visual medium is that sufferers had frequently commented on the difficulty of articulating the appalling horror of the panic-attack, the experience that sows the seeds of agoraphobic avoidance. As we have seen, respondents draw on a wide range of metaphors in their attempts to verbalise the experience, likening it, for example, to such diverse phenomena as 'drowning', or 'receiving electric shocks'. Given such barriers to communicating the experience, I was interested to see how the producers of these videos might attempt to re-present panic visually, and whether or not the outcome would be meaningful for agoraphobic viewers.

As for my selection of these *particular* products, video A seemed an obvious choice, for reasons such as familiarity of location, accent etc. to members of the group. Additionally, I rightly suspected that the group would be keen to engage with a text produced by individuals described by one member of the audience as 'people like us' (Ruth). Video P was chosen because it came most highly recommended of those currently available, by a recent newsletter produced by *No*

Panic, the UK wide self-help organisation for panic sufferers. It advocates a form of cognitive behavioural therapy (CBT) which, I argue, sits firmly within, and does not problematise, the Cartesian tradition of selfhood.

Re-presenting Panic

I want now to open the discussion of the self-help texts by examining their portrayals of the experience of panic. Perhaps unsurprisingly, given the account of spaces of consumption in chapter 3, 'shopping scenes' form a key aspect of both videos' portrayal of the agoraphobic's experience. We are already familiar with the significance of the *agora,* or *market* place, and social spaces of this particular kind often present sufferers with the greatest difficulties. The opening scene of video A (the 'amateur' production) is Prince's Street in Edinburgh, one of the largest and best known shopping sites in the city centre. This text begins by positioning the viewer around the halfway point of this street, looking down a stretch of pavement along which crowds of people move, at various speeds, in both directions. In the centre of our view, a woman walks towards us, clutching her bag tightly in front of her. Her eyes dart around nervously, then she stops, and stands with her hands pressed to the sides of her head, covering her ears; she appears desperately uncomfortable. A steady, rhythmic, pulsing sound can be heard, increasingly loudly, in the background, and seems to simulate ultra-awareness of a rapidly beating heart. During these opening seconds, the following three accounts of panic are overlaid.

> I feel faint, sick and dizzy, and I come out in a cold sweat.

> My heart starts hammering, my chest feels tight, I can't breathe, and I feel as if I'm having a heart attack. The noise of the traffic seems to get louder and louder, and I feel more and more frantic.

> I begin to feel absolutely rooted to the spot, and I just can't move at all. I feel as if there's an iron band tightening round my head.

The scene then changes to an aisle in a supermarket. It is brightly lit, even gaudy, and the light pulses in time with the repetitive beating sound. As we move forward, slowly and unsteadily, the aisle appears increasingly tunnel like, and never quite comes into focus. Objects on the shelves become closer, then move away, but their outlines remain blurred. Our perspective then alters, unexpectedly and quickly, as the camera pans upward, towards the light, then lurches down. The ground beneath us glares, blindingly, and our view slowly and shakily lifts to eye level. The effect is one of disorientation, and this is clearly the intent. We hear Sheila's voice:

> Going to the supermarket is an everyday experience for most people, but to an agoraphobic, it is a *real* trauma. The stimulus is just too much for the eyes to cope with, and the fluorescent lighting, bouncing off the already tiled floor, causes a feeling of anxiety, if not *extreme* panic.

The shopping scene in video P (the 'professional' production) opens with 'Sharon', seated outdoors, amongst greenery, in what is presumably her garden. Her account begins:

> And then I was in a shopping centre, with my sister, the one day, and I think that's when I really got a full-blown, out-of-control, panic attack, where I was standing in the shop with her, and just, all of a sudden, my world just [pause], the world started spinning, and I just thought I was going totally out of control. [The scene changes to the inside of a shopping mall. The camera is out of focus, and moves with a rapid freeze frame action in a circular motion. Sharon's voice continues throughout.] My legs started going weak, um, I couldn't focus on anything, um, I really started shaking, almost [pause] hot and cold chills going down me, um, rapid heart beat, and I just actually thought I was going to collapse right there. Terrible nausea, and em, I remember my sister looking at me and saying, you know, you are absolutely white as a ghost!

This scene is accompanied by dissonant, drawn out notes played, perhaps, on an organ. It begins erratically, then alternately ascends and descends an atonal scale, which is obtusely reminiscent of an alarm, or siren. The sound connotes confusion, disorientation, and the irregular interval between each note is suggestive of strangeness, unfamiliarity. What we actually see is an excess of bright lights and reflective surfaces, a predominance of glimmering whiteness that is decidedly not easy on the eye. The camera moves in on signs, perhaps advertisements, which shift in and out of focus and are impossible to read. People, too, move in and out of focus, and the scene appears disjointed, unpredictable. The viewer's perspective appears odd, and changes without warning; side to side movement creates a sense of being off balance, and there is a jerky quality to the camera movement, suggesting a lack of continuity. The place looks thoroughly menacing, and the viewer has no control over the way visual information is gathered and presented.

Perhaps the first thing one would note about the shopping scenes would be the striking similarities between the visual portrayals of panic by both videos. Although it is apparent, and predictable, that there is greater technical expertise available to the producers of video P, clearly both are aiming to produce similar effects. Both intend to create a sense of disorientation, confusion, lack of focus, and of the overwhelming fear that would accompany the loss of control over one's senses, one's relations with the environment and ultimately, of one's self.[5] This element of 'loss of control' can be seen to constitute the central theme of both portrayals; all aspects of these scenes, visual, verbal, and otherwise, are primarily geared toward communicating this state of being to the audience. Panic, the viewer learns, entails loss of control to a terrifying degree, not only over oneself, but crucially, also in relation to the environment.

It is my contention that this portrayal supports the interpretation of agoraphobic anxiety advanced by this book as a *boundary crisis* – a term also used by Kirby (1996) in her partially autobiographical account of 'post-traumatic vertigo'. The loss of *self*-control relates to a breakdown in the normally taken-for-granted relation between, not mind and body, but self and space, a subjective experience for which the Cartesian worldview does not allow. To warrant this interpretation, first consider the images in isolation. The presentation of visual information in both shopping scenes is strongly suggestive of lack of agency on the part of the protagonist. Think of the scene in video A where the supermarket aisle closes in on the subject, or where the ground lurches upwards. In video P objects are shown to move rapidly closer or to become unexpectedly bright. One could interpret these images as suggesting that the balance of power in the relation between person and place has shifted from the former to the latter. This is to say that, for the agoraphobic, the environment itself takes on a kind of agency. It loses its static, predictable 'thingness', and this lack of assumed continuity renders it threatening and dangerous. How can you feel secure in a place where the visual information it presents you with virtually *assaults* your sense of sight, and thus, your sense of self?

The notion that the self is under threat is supported by the linguistic accounts that accompany these images, as is the earlier suggestion that we can identify a central theme. In video P, Sharon characterizes her first panic attack as 'full blown, *out of control*', and describes her environs thus; 'my world just, *the world started spinning*'. The effect of this sensation? 'I just thought *I was going totally out of control*'. Sharon moves between descriptions of apparently 'internal' and 'external' happenings with manifest ease. It is 'the world' that spins, but she herself that is 'out of control', and I take this to be indicative of ambiguity, fluidity, in the boundary between the two. Her characterization suggests that her sense of herself is inseparable from her sense of her environment, and is thus in tune with the relational construction of subjectivity put forward by feminist and post-structuralist theorists such as Judith Butler (1990: 1993).[6] What Sharon goes on to say, however, seems to focus entirely on *internal* bodily sensations, implying more stability of boundaries, more autonomy, than my interpretation suggests. By her own account, these things are happening *to*, rather than around her.

If, however, we take post-structuralist feminist approaches seriously, the sensations Sharon describes need not be understood so exclusively, as definitively within *or* without, the limits of the flesh, as an attribute *either* of subject *or* object. Theorists who are critical of dualistic conceptions of subjectivity, rather than those with a Cartesian perspective, can thus help articulate the experience of agoraphobic panic, which *exceeds* bodily limitations, and corrodes the security one's boundaries afford. Consider, for example, the embodied experience of shakiness, nausea, 'hot and cold chills': these phenomena are symptomatic of a perceived disruption in the barrier between internal and external space. They are the subjective sensations of corporeal *breaches*, im*mediate* signs of permeability that are anathema to the stable subject of modernity.

In video A, participants present experiential accounts that are similarly suggestive of a confusion in self/space relations. While some of the sensations described are apparently internal – 'my heart starts hammering', some apparently external – 'the noise of the traffic seems to get louder and louder', others are concerned with the interactions *between* internal and external realms. To be 'rooted to the spot', for example, and to 'have an iron band tight round my head', is to make use of metaphors arising from person/place relations. Other seemingly straightforward descriptions also turn out to be more ambiguous on closer inspection. The statement, '*I come out* in a cold sweat', clearly implies a breach of the internal/external barrier, and Kathy's choice of words suggests that she her*self* is exposed in the process. Similarly, to be 'faint, sick and dizzy' are all symptoms that disrupt the felt boundaries of the self. Lucy Yardley has paid particularly close attention to experiential accounts similar to these in developing her 'ecological-constructionist' analysis of disorientation and dizziness. Yardley's analysis is, arguably, also applicable to agoraphobic anxiety. Here, the subject also experiences disorientating phenomena, and doubts their ability to respond appropriately to their environment (Yardley 1997: 117), to maintain *self*-control.

The importance of Yardley's account for agoraphobia lies mainly with its focus on *boundaries*, and its refusal to work within reductionist frameworks dominant in contemporary discourses of health and illness.[7] In her own words,

> the ecological account characterizes disorientation as arising at the *interface* between the individual and their environment … [it places] the emphasis on dynamic relations rather than essential causes … This is no simple psycho-somatic or somato-psychic relationship, but a profound and intricate interactive bond between the material and discursive aspects of disorientation, which creates and reproduces the condition and preconditions, meaning and implications of disorientation (Yardley 1994: 118-9).

The ecological-constructionist view presents us with a means of understanding the subject *in context*. And, as the shopping scenes clearly demonstrate, context is intrinsic to understanding the nature of agoraphobic experience.

The women with whom I watched these scenes were clearly impressed by the faithfulness of their representations. The following exchange came immediately after the shopping scene in video P, which we viewed first.

> Betty – That's what it feels like to me, that is what it feels like [pause] but it feels like that to me out in the street as well.

> Jane – Like things go out of focus.

> Joyce – So d'you think that would be quite a useful thing, if I was to show that to people, and say that's what a panic attack's like [pause] is that fair?

> Ruth – Oh aye, that *is* what a panic attack's like.

Their response to the scene in video A is similarly enthusiastic, with comments like 'that's what a panic attack's like, like a switch getting turned on' (Jane), and 'that's the truth, that's how it really happens' (Ruth). It is reasonable to claim that, for these sufferers at least, the shopping scenes managed to capture, and communicate something of their experience. Inevitably the portrayals are particular and partial, in that they try to give visual sense to *sufferers'* (as opposed to medical practitioners') descriptions of anxious experience. It is perhaps for this reason that the representation is felt to be 'authentic', that it is somehow faithful to the spirit of panic, whose picture it attempts to paint. There is, as of yet, no attempt in the texts to *explain*, or draw conclusions about these phenomena, and the face of panic remains recognizable to those who suffer its presence in their lives.

So far, I have argued that there is a dominant theme shaping both of these shopping scenes, namely, the severe loss of control suffered by the agoraphobic subject. I have further suggested that this loss of control can be most constructively understood in (*non*-dualistic) terms of a breakdown in the relations between person and place. I now want to consider the format of the texts as a whole, in order to place these portrayals of agoraphobic selves in relation to the Cartesian, *masculine* ideal of normality against which they are set.

(Ab)normal Narratives

The remainder of video A consists largely of eight narrative accounts of agoraphobia that are presented by the sufferers in their own homes. All are introduced by first names, and all but one are women. The pieces are edited from interviews/discussions with Sheila, who is herself a sufferer, and was the organizer of the self-help group at the time the video was made. Sheila narrates throughout, occasionally commenting on sufferers' accounts, or asking encouraging questions. It is Sheila who introduces the sole non-phobic respondent, whose contribution will be assessed below.

Video P begins with a male voice describing a first panic attack, and his subsequent fear of driving. The scene switches between 'Wayne' in his living room, and footage of a night driving scene, while he describes how panic would 'take over', and how he thought he 'might die'. Enough information is given by this sufferer for us to deduce that he was employed as a driver, though his first name is the only written text to appear on the screen. Wayne's piece lasts for around 90 seconds, and is followed by the 40 second piece involving Sharon which is described above. Immediately afterwards the scene changes to the office of Dr. Lisa Lampe, who, the caption informs us, is a consultant psychiatrist. Lampe proceeds to reorganize and reiterate much of what Sharon has said, listing the symptoms of panic in an orderly manner. Professor Ron Rapee takes a similar role when he appears on screen around 30 seconds later. Rapee, a 'professor of psychology', offers another dispassionate account of panic episodes, and concludes definitively

> if a person starts to associate those panic attacks with certain areas, and because of that they start to avoid doing certain things, or going to certain places, then we would call that agoraphobia.

The four characters involved in video P have then all been introduced within the first few minutes of the video. As we will see, the layout of their introductory pieces sets the tone for the rest of the video. Sharon and Wayne alternately describe their experience, and the professionals interject to retell their stories in an authoritative fashion, supplementing them with additional, technical language and information. In 'translating' individual accounts in this way, Lampe and Rapee could be said to undermine the phobic contributors' authority. They also move beyond the particularity of the personal narrative, toward a more generalized account, arguably a meta-narrative, of agoraphobia.

The authoritative identities of Lampe and Rapee are maintained throughout the text in a number of ways. First, both are introduced with their full professional titles, which has the effect of fore-grounding their institutional, rather than personal roles from the outset. This effect is perpetuated by presenting both as entirely static throughout the video. That they remain seated in what are presumably their own offices creates the impression of professional consultation in an institutional context. Additionally, both are presented in formal dress; Lampe and Rapee are seen in smart office-wear, as opposed to Wayne's T-shirt, jeans and overalls, and Sharon's loose, patterned dresses. The manner and style of their delivery is also contributive. Both use calm, measured tones, and their use of language tends to be authoritative. The use of medical and semi-technical terms serves to enhance the impression of expertise, as does the occasional use of statistics – 'as many as thirty five, or more, per cent of the population will have a panic attack, at some point in their lives'. The style of the professionals' delivery is then largely pedagogical.

In these various ways, the impression is created that Lampe and Rapee are providing unbiased information for the enlightenment of their audience. This information is never presented as opinion, and we are given little encouragement to doubt the veracity of their accounts. Neither Sharon nor Wayne questions the authority of any statement, and so an impression of consensus is created. Lampe and Rapee are also clearly at a distance from the subject(s) they are discussing (quite literally – 'professionals' and 'sufferers' never appear on screen at the same time). They are never *at pains* to communicate the reality of the disorder, so their contributions are less emphatic, less intense, even incompatible with those of Sharon and Wayne. There is none of the immediacy that often characterizes sufferers' accounts; rather, the professionals' approach to agoraphobia is 'objective' and 'rational', directed toward bringing an unruly body under discursive control. This approach privileges and re-emphasizes the dominant (masculine) side of the hierarchic dichotomies referred to above.

For the producers of the amateur, as well as the professional production, there is a specifiable, physical, reason and explanation for agoraphobics' 'abnormal' loss of control. *Not all in the Mind* posits a balance dysfunction as the root-cause of sufferers' disorientation and anxiety. Their inability to maintain a 'normal' level of self-control is, consequently, *not their fault*. This physical, rather than mental, attribution of the problem is an important point since, as Yardley emphasizes, there are 'moral implications' to this far from neutral issue.

> [D]eviations from normal behaviour for which no somatic explanation is provided inevitably carry a social penalty, since psychological problems are considered less 'real' and more blameworthy than physical defects (1994: 115).

It is then very much in the sufferers' interests to demonstrate that their 'abnormal' behaviour is a *reasonable* response to a physical problem. While adopting this viewpoint, the manner of sufferers' representation of rational (masculine), biologistic medical discourse arguably demonstrates resistance to its hegemony.

In video A Dr. David Weeks is presented in a manner that foregrounds his institutional identity, using largely similar methods to those of video P. However, the surrounding contributions from non-professional participants serve to *contextualize* his expertise. He is introduced as someone who is conducting research on sufferers' behalf, and is called upon to present findings that concur with the group's view of agoraphobia's links with problems with balance. Weeks is being used to verify, give the stamp of authority to, what the sufferers' themselves already knew. He begins;

> We discovered that agoraphobic balance problems are genuine and real ... this condition is not purely psychological, there's a heavy, physical, biological, neurological component to it, that has been only marginally suspected in the past, but never actually proven. Now it's been proven.

This contribution is clearly intended to be authoritative, but we can see that the agoraphobic women in control of this text have opted to include the professional's account in order to add credence to their own theories about their disorder. They have put themselves in the position of being able to choose who to include and who to exclude, in order to construct their own version of agoraphobia, perhaps realizing that 'discourse is the power which is to be seized' (Foucault 1984: 110).

I have already suggested that the privileging of the masculine constitutes a recognizable theme in self-help discourses. In the following section I want to explore the (implicit and explicit) depictions of gender in more detail, focusing particularly on the implied association between masculinity and endurance, femininity and fear.

Framing Fear as Feminine

> If it happens to you first of all when you're in your teens and you go to see the doctor, the doctor will say, "it's your age, it's adolescence. When you're a bit older, you'll get over that problem". When you're in your twenties, they say "once you're married, and have children, and have plenty to do, you'll forget all about that silly business". And then, you get to the menopause, and they say "once the menopause is over, you'll be much better". What happens then? What do you wait for then, 'til you die, will you be better then? (Sheila).

This quotation, which brings video A to a close, suggests that some sufferers are more than familiar with the practice of attributing their fearful condition to their sex and/or expected gender role behaviour. For many, agoraphobia is a woman's problem, or even a 'housewife's disease' (da Costa Meyer 1996: 150) – and (thus?) an *irrational* and thoroughly 'silly' affair.

The women with whom I watched this video responded with much hilarity to these closing remarks, although their cognizant laughter was perhaps tinged with the *bitterness* of remembrance. They too had experience of this kind of dismissal and a refusal to take their 'trivial', womanly problems seriously.

The assumption that agoraphobia is a feminine affair is often largely unspoken. It is Bill, the sole male agoraphobic contributor to video A, who comments explicitly on this gendered aspect of the disorder. He states:

> Quite frankly, in 25 years of suffering from agoraphobia, I have never met another male who has said that he is agoraphobic, and yet, I am told, that there are a number of men, who do suffer, and em, I feel that most [pause] that probably says a lot. Men have to think about their image, the macho image [so it is difficult] to admit to suffering from something which a lot of people simply see as being a weakness.

Bill suggests that men are more likely to resist the agoraphobic appellation then women, but *not* that they are less likely to have the same anxious experiences. While his claim is clearly contentious, that a sufferer perceives this to be the case is in itself significant.[8] Bill highlights the notion that agoraphobia is widely perceived to be a peculiarly *feminine* weakness. Given the male-stream western tendency to belittle all that is *not*-masculine, men may well be reluctant to associate themselves with this denigrated state of being-in-the-world. Bill's comments on his solitary experience as a male agoraphobic reveal what he sees as identifiable, understandable *reasons* why men will refuse this particular subject positioning.

Wayne, commenting on his own disorder in video P, points to personal difficulties he encountered in admitting he was agoraphobic, difficulties we can perhaps relate to Bill's notion of the male 'macho image'; 'at first, my pride wouldn't let me do it [pause]. I'm not in that basket, if you know what I mean [laughs], that's not what's wrong with me at all'. (Although Sharon expressed similar hesitancy, suggesting it is not only men who would wish to distance themselves from 'feminine' characteristics.) Overall, however, video P creates an entirely different impression of the gendered distribution of agoraphobia from video A. The inclusion of one male and one female sufferer suggests a gender balance, and there is no mention that this is not in fact representative of the actual gendered distribution of agoraphobia. One might then wonder if the producers were concerned *not* to represent agoraphobia and panic attacks as 'women's problems', perhaps even to disrupt conceptions of fear as stereotypically feminine.

This notion can be supported by the fact that the text's portrayal of Wayne is often framed in terms of traditional constructions of masculinity. For example, his role is often active, and he is repeatedly seen performing tasks associated with paid employment outside the home. Wayne works on an outdoor pool, files papers and answers the telephone in his office, and attends to the mechanics of his racing car in his garage. These performances are markedly different from those of Sharon, who behaves 'similarly' only to the extent that she too conforms to her gender-role stereotype.

Sharon is shown standing over a kitchen sink, looking out of the window, or peering through a door that she is holding slightly ajar, while her voice-over describes her inability to leave home. When we do see Sharon out of doors, she is driving her car to collect her children from school, standing in a queue in a post office, and entering a shopping centre, all activities associated in some way with domesticity. It is thus Sharon's role as a housewife that is fore-grounded by this text, and this clearly has implications for the way her contribution will be received by different audiences.

The visual representations of gender in video P might then be seen to challenge assumptions about the feminine nature of fear, by associating it with displays of masculine, even *macho*, imagery (racing car mechanics, DIY, etc.) and not simply with (feminine coded) domestic chores. However, the potential power of this association is undermined by the text's consistent refusal to trouble gendered dichotomies in any other way, a tactic which effectively amounts to complicity in their hegemony. A greater degree of authority and credibility will thus continue to be associated with a male, rather than female voice.

In video P, it is in fact Wayne's voice that opens and closes the text, and it is Wayne who is given the task, along with the professionals, of conveying information about new self-help techniques, and demonstrating how they are to be carried out. These factors strengthen the sense of authority created by his 'manly' activities, adding the weight of cultural capital to his masterly, masculine presence. Although it is Sharon who is experientially best placed to speak to agoraphobia sufferers, her contribution is denigrated, designated *less* than that of the male by framing it in terms of culturally undervalued domestic work. Whether deliberate or not, the gendered selves represented by this text are echoes of those masculinist stereotypes predominant in media discourse and the wider socio-cultural realm of text production. The portrayal of gender is always the Same, and agoraphobic fear is still largely a 'feminine' affair. Even if there are occasional aberrant males displaying similar symptoms, they must have the self-same ideal. In line with the privileging of masculinity we find throughout the texts, the advocated response to 'aberration' is *aspiration* – striving to be yet *more*-of-the-Same; more manly, more in control, and above all, more rational.[9]

From Emotional Woman to Rational Man

At various stages throughout the texts, sufferers are advised to 'rationalize their thoughts', and to *practice* 'rational thinking'. Agoraphobic fears are, on this account, *physical* in origin, and nothing that a good dose of rationality won't cure. They arise when subjects respond irrationally to physical symptoms, and can be brought within their conscious control.[10] As Rapee states in video P:

> What happens with people with panic attacks is they get these feelings, these physical feelings, that, that begin their anxious feeling. They then start to think of these feelings as being something terrible. They've learnt across their lives that these physical feelings are bad, "I shouldn't have them", so they immediately start to be anxious ... The important thing for people with panic attacks to learn is that these are two separate things. There're physical feelings, and then there's your interpretation, your emotional response to those things. [Rapee advocates mentally separating these sensations, in order to see them much more objectively.]

It is important to be clear about the content of this statement. It contains the *definition* of panic, and is presented as an entirely factual account. And yet this account is contentious, to say the least. Rapee effectively states that fear is a *response* to adrenaline, by no means an obvious or 'common sense' conclusion, and one that many sufferers would dispute (see below). What we have here is an account that constructs the agoraphobic as exhibiting an irrational loss of control in the face of an unruly body. There is an assumption that nothing can be done to prevent the physical occurrences, that the body is beyond hope or restraint. However, one can learn, be taught, to control one's *responses* to its behaviour, thereby re-establishing the body in its proper place in the mind's dominion.

Given that this dualistic model informs the professional view of the agoraphobic subject, we can begin to see the origin of what I term the 'bureaucratic' approach to her treatment, one that is quite literally a case of mind over matter. Lampe and Rapee effectively advocate the development of a good *management* strategy, to enact their positivistic and quantitative approach to 'treatment'. First and foremost, the subject must accept their designation, and be fully *interpellated* into the agoraphobic subject position. Wayne himself makes this point clear; 'until such time [pause] as you are diagnosed, and *you accept, that that's what's wrong with you* [pause] you won't start getting better' (emphasis added). He then adds, in terms one might describe as 'macho', that it is also necessary to 'have the fight or the will to beat it'. Having satisfied these conditions, the subject can embark on a programme of 'cognitive restructuring', in which 'self monitoring' plays a crucial role. This process, the *monitoring* of the self, is aided by the provision of specially designed forms that are to be carried about the person at all times. This tactic enables panic symptoms to be recorded accurately and efficiently, and thus be objectified, 'as soon as they've finished having an attack' (Rapee). In the appropriate boxes, sufferers should note details such as the time and the duration of the panic, as well as the intensity, 'on a 0-8 scale'. Gathering 'evidence' about, and quantifying symptoms, is intended to encourage the development of less 'extreme', more 'realistic' thoughts. In addition, sufferers should 'learn the right rate' for breathing, which is held to *be* 'three seconds in, three seconds out', though appropriate adjustments can be made (with the aid of a second hand on a watch or alarm) if this is not found satisfactory.

Until now, I have clearly chosen to focus on vocabulary that emphasizes the scientistic nature of the regulatory discourse. However, much of this terminology is actually couched in more 'ordinary', colloquial phrases, and delivered in a conversational style. For example, Rapee describes the self-monitoring form as 'very nice as it comes as a little pocket sized [form], so that people can always carry a bunch' with them. The effect of this 'conversationalization' is, as Fairclough (1995: 10) has argued, to *naturalize* the content of the speech, to suggest that these are run-of-the-mill, *common sense* notions shared by all of 'us'. In fact, Rapee is often more explicit in his attempts to normalize his normative views; he describes the treatments he advocates as 'practical common sense ones'. He further attempts to naturalize *biologistic* aspects of his model of the self. For example, Rapee suggests that when a panic episode begins, the sufferer should try thinking

> that's a good thing that my heart's pumping blood round my body, and helps me get ready to fight. It's something very natural and normal, it's simply happening at the wrong time.

One should, then, accept that the body is behaving 'naturally', but only insofar as we can *use* this information to enculturate our volatility (knowledge is power, after all). Our bodies can be *civilized* by the reach of the mind, and rational, realistic thinking should be practised until it becomes, ironically, 'almost second nature' (Wayne). Sharon too advocates the power of mind over body, reason over emotion: 'I still, to this day, have to do um, practice rational thinking, and I'll probably have to do it maybe for the rest of my life'.

The women with whom I watched video P were not always convinced by its 'common sense' characterization of agoraphobia. I ought to admit that they could have picked up on the fact that I personally was more favourably disposed toward the amateur production. Besides this, however, it is difficult to foster a sense of sharing a life-world with the characters involved, not least, because of the strangeness of their surroundings. On a winter's day in a Scottish housing estate, it's difficult to empathize with someone basking in the Australian sunshine by the side of their outdoor pool. There may well be other mitigating factors, but it remains significant that members of this audience were critical of some of the assumptions underlying this version of agoraphobia, and that they drew on their own experience to validate these criticisms. Iris, for example, commented on the nature of panic thus:

> I think sometimes the physical feelings are the result, that it's the other way round [pause] the emotional feelings that you have lead to the physical things.

Jane tentatively agrees, then states,

> I mean [pause] the effect's there, the whole thing, before I start thinking, my fingers are tingling, or my heart's starting to beat a bit fast, or I feel dizzy. It's [pause] the whole lot comes [pause] phwoooh [this is accompanied by hand gestures, to simulate something rushing toward her face], without any reason.

Sufferers' accounts suggest that they simply do not experience this distinction between physical and mental symptoms in the way the above examples from professional discourse suggest. Their vocabulary and metaphors of panic point rather to a concern with the relationship between themselves (as deeply embodied) and their surroundings; that is to say, with their immediate psycho-corporeal geographies. As one would expect, Cartesian conceptualizations of subjectivity are not adequate to the task of articulating the experience of dynamically constituted subjects.

The reduction of disorientation [or agoraphobic anxiety] to purely physical or wholly psychological factors presents many individuals with an artificial forced choice between the somatic and psychic models, with the result that *important aspects of their experience remain unexplained and untreated* (Yardley 1997: 114, emphasis added).

Moreover, sufferers' difficulties are compounded by this impossibility of fully articulating their experience. '[T]hey are forced to resort to the unsatisfactory discourses available to them in order to (mis)represent their problem to themselves and others' (Yardley 1997: 115). This does not, however, mean that the *therapies* reliant on such 'unsatisfactory discourses' are completely without merit.

Clearly, it must be acknowledged that the cognitive behavioural therapy (CBT) being advocated here does in fact 'work' for some agoraphobia sufferers, in the crucial sense of giving them a hold on their disorder, and thus making their everyday lives more tolerable, less infused with panic. Despite my obviously critical analysis of the ideological underpinnings of CBT, such apparently academic matters need not necessarily be a concern for the subject of self-help whose anxiety can be partially soothed and steadied, *slowed down* by counting, controlling measures.

However, I want now to draw attention to potentially *substantial* problems associated with an over reliance on this particular model of the self, one largely divorced from the all-important social context. I have suggested that the supposed 'universality' of the self represented in these texts is challenged by the embodied experience of agoraphobic women, thus revealing a gap between (at least) two contesting conceptions of self. However, despite the texts' representation of agoraphobic women as problematizing the Cartesian view of the self, the professionals' response to their experience is to continue to privilege and re-emphasize this *Same* (masculine) notion of self. The superiority of this model is uncritically assumed, and sufferers are left in no doubt that they ought to (com)press themselves into an ill-fitting Cartesian mould, a restrictive mould which, I would tentatively suggest, may be partially responsible for their dreadful difficulties in the first place.

Conclusion

This chapter has advanced a feminist interpretation of self-help texts for agoraphobia sufferers, stressing the importance of rendering underlying assumptions about the nature of selves explicit. I have argued that it is a *Cartesian* notion of self that is uncritically assumed by the texts in question, and that this has important implications for the treatment of agoraphobic women. The point to be emphasized is that therapies constructed around notions of self to which agoraphobic women cannot easily relate will have *limited* effectiveness. Sufferers most likely do not feel, may never have felt, autonomous, rational or firmly bounded as subjects, surrounded by yet separate from a world of isolable objects. Within dualistic discourse, however, there is little room for alternatives, and the agoraphobic woman can be seen only as somehow lacking and inadequate; *like*, but less than, the rational (male) ideal. As Rose points out, '[d]ualisms maintain order by offering only two positions, both of which are constituted around a single term – the masculine Same' (Rose 1993: 82). There is as of yet no conceptual position to comfortably house the excessive experience, the (to borrow from Irigaray [1985: 112]) *ex-orbitant* bodies and voices of agoraphobic women.

My intention is not, however, to argue that we should or even could ever entirely reject this Cartesian model or the dualisms at its heart. Dualistic thinking is obviously firmly entrenched within our language and worldviews, utterly pervasive throughout all aspects of society. That it can operate to stabilize troubled identities, more or less temporarily, is not in doubt. And in fact, rather than celebrating their 'difference', the person who is 'ill' will most likely want nothing more than to 'fit in', to conform with their society's expectations of normality and rationality. Yet this is not to say that dualisms should not be problematized, that we should not question exactly what *kind* of society, what kind of selves, they are helping to produce and maintain.

This discussion has shown that unquestioning and uncritical approaches to subjectivities are severely *limited*, incapable of conceptualizing or containing the full extent of agoraphobic being-in-the-world, and that in fact, the results are always the Same. I want to suggest that, by analytically probing predominant conceptions of normality, we might create potential spaces for alternative models of self to arise, models that allow difference in terms other than deviation, and to which '*ill*-defined' women could more comfortably relate.[11]

I am *not* suggesting that such strategies could somehow alter the phenomenology of fear, that the actual lived experience of panic and anxiety could be rendered more tolerable. Rather, that the troubled identities of agoraphobic women are not *so* far removed from an acceptable (feminist) 'ideal' of normality that they should be beyond representation. Indeed, respondents' inspired use of metaphor, and the imaginative visual portrayals of panic in self-help videos are indicative of the kind of communicative work that must be done. We initially require a 'vocabulary' with which to articulate such experience, and this must be the practical goal of a strategy to make discursive space for *integrated* corporeo-conceptual involvement with the world. This strategy would also be Utopian, however, in that it reaches out toward the *no-place* beyond dualistic thinking, speaking and writing.

It has long been the contention of many feminists that language *matters*, and one's ability to mobilize discourse to communicate with others is of fundamental importance for innumerable aspects of our lives, not least, for our mental health. The personal narratives composed by agoraphobic women draw on metaphoric language in particular ways, to communicate how their experience consumes and confuses their senses.[12] They are 'drowned' or 'suffocated', 'shaken' and 'shocked', 'frozen' in both time and space. Their experience is *deeply* embodied, but in a way that blasts through dualist borders.

I want to suggest that the language used by these women to express their experience is at times an attempt to transcend the dualisms implied by, and inherent to, our specular symbolic order (Irigaray 1985). Consequently, it may not always be helpful therapeutically to (op)press this experience back into dualistic discourse, to interpret it strictly, for example, in terms *either* of mind *or* of body.[13]

Perhaps then, feminist theorists and therapists should acknowledge and encourage the use of spatial and other metaphors to *ground* such boundless and fearful exposure, to create a platform of sorts on which it can be articulated and, therefore, granted legitimate existence. We might constructively develop existing elements of our vocabulary that do not insist on the mind/body split, and which therefore resist complicity with masculinist dualistic thought. Boundary breaking speech acts can be literally transgressive. By developing feminist and therapeutic language games[14] with a more integrated, holistic approach to selves and space, we may be more adequately equipped to faithfully represent agoraphobic, and perhaps *Other*, experiences, to provide alternative-self help. Perhaps now, with this re-weaving of feminist, phenomenological and agoraphobic narratives, we might begin to see different patterns emerging from the tangled threads of women's lives.

Notes

[1] Val Plumwood characterizes dualistic thinking and its place in western philosophy in terms of a 'logic of colonization', describing the concept of dualism as 'the construction of a devalued and sharply demarcated sphere of otherness [which] results from a certain kind of *denied dependency* on a subordinated other' (1993: 41, emphasis added).

[2] For examples of self-help *books* for agoraphobics, see Savage (1987), Marks (1980) and Vines (1987).

[3] See Parker and Burman (1993) for an overview and discussion of alternative approaches to discourse analysis, and of associated problems.

[4] Duley (1997) has adopted Fairclough's method to provide an exemplary account of the construction of New Zealand national identity through analysis of a series of television broadcasts.

[5] Although see Yardley (1997: 123) for a discussion of cultural or drug induced alternatives, where 'loss of control' is positively welcomed. She writes that 'there are clearly important differences between disorientation which is deliberately self-induced, whether by dance, drugs or dazzling displays, and sudden attacks of vertigo and dizziness which were neither anticipated nor desired'.

[6] See also, for example, Lorraine's (1999) exploration of Irigaray and Deleuze on subjectivity, and Battersby (1998).

[7] See Lupton (1994), and Jones and Porter (1998), especially contributions by Armstrong, Osborne and Rose.

[8] Dianne Chambless has presented research findings that may be taken to support Bill's claim, suggesting that agoraphobia in men may often be masked by substance abuse. She writes, '[g]iven the prevalence of alcohol in Western society, it may be expected, then, that large numbers of male agoraphobics are to be found, not in phobia societies and psychiatric clinics for phobia treatment, but in bars, Alcoholics Anonymous, and alcoholism treatment centers' (Chambless 1982: 5). However, her findings have yet to be replicated, and the vast

majority of research on agoraphobia continues to point to a significant sex difference in distribution (for example Marks 1997, Mathews et al. 1981, McNally 1994).

[9] Gillian Rose (1993: 75) provides a useful account of the phallocentric nature of dualistic thought. In her words, the two sides of dualisms 'are not two discrete alternatives, because the feminized side is defined in relation to the masculine. The two sides are not oppositions between two unrelated terms ... Rather, this is a field of knowledge divided between two related terms. Woman is described in terms of Man, as the Other of the Same'. This 'field of the Same' 'cannot admit radical difference from itself'. According to this masculine symbolic order, women are both *partially* Other and the Same, but *fully* neither. They are thus always already unstable.

[10] For discussion of feminist critiques of rationality, see Hekman (1992).

[11] For alternative criticism of 'normal' conceptions of embodiment, see Dyck (1995) and Park *et al.* (1998).

[12] See Kearns (1997) for discussion of narrative and metaphor in relation to health and illness.

[13] Where such dualistic splits are desirable and helpful, we need, at the very least, to explicitly recognize that they are being employed for strategic purposes, and acknowledge that the split may not seem to accord with individuals' embodied experience of the world. It is not a naturally occurring boundary, but rather *a* particular way of conceptualizing identity, and one that will have both positive and negative effects. There are other possibilities.

[14] For a feminist interpretation of the Wittgensteinian concept of language games, see Davidson and Smith (1998).

Conclusion

Re/solution of Spatial Identities

In bringing this book to a close, I aim to draw its diverse experiential and theoretical threads together, and to underline its various insights into the experience and meanings of agoraphobia. I claim that the book's findings are not only of potential significance for the disorder's sufferers, but that they constitute a significant contribution to socio-geographical understandings of the mutually constitutive nature of selves and spaces. Before outlining the form and content of this contribution, I will present a brief reminder of the project's background and beginnings, and the process of its development towards this final form.

From the outset of this work, I aimed to be clear about the emotional, intellectual and political roots from which it grew. Its substantially feminist and phenomenological approach was shown to have evolved in response to a perceived predominance of bio-medical and psychological interpretations and explanations of what is evidently *felt* by its mainly female sufferers to be a socially and spatially mediated disorder. In an attempt to address some of the gaps revealed by this initial (and on-going) review of relevant literature, an emphasis was placed on the individual (and later, collective) voices and narratives of its mostly female sufferers. By placing such first-hand experiential accounts of the disorder centre stage, I hoped to make space for the largely neglected stories of agoraphobic women to be told and heard. Such a strategy was thus meant to prioritize the views and concerns of sufferers themselves, and to allow the research project to develop in a responsive manner. Further, by presenting in-depth inquiries into individual situations – and so *situated* selves – it aimed to emphasize the importance for understandings of agoraphobia of recognizing its immediate (embodied) and wider social and spatial contexts. In these ways, it was hoped that a new, more sympathetic and inclusive account of agoraphobia would be allowed to emerge. Actually realizing these aims, however, proved to be a complex and ongoing process.

In chapter 2, I explained how self-help groups came to occupy a more central place in this research, to mirror their centrality in sufferers' lives. The changing and increasingly intricate nature of my involvement with groups was in fact to provide much food for thought. Admittedly contrary to my expectations and/or preconceptions, I quickly learned that the members of self-help groups were often willing and more than able to 'manage' my role within their group. The manner of my participation was thus often shaped in response to members' more or less explicit attempts to position me in certain ways. With one group in particular, this positioning and enactment of a 'communicative distance' between us was 'measured' and articulated via the medium of humour. Group members' expressions of humour not only emphasized the extent to which they shared certain embodied knowledge and experience – and so, I argued, a 'form of life' – it also accentuated the extent to which they saw me (and at first, endeavoured to keep me) in the role of 'outsider'. Humour can clearly be used to maintain, shift, or break down boundaries between individuals and groups. And in fact, as chapter 2 demonstrated, the communicative ('humorous') distance between myself and the self-help group was open to question, and decreased over time as the women became willing to 'let me in on' their jokes. Respondents' jokes were thus shown to have an important, indeed revolutionary role in the negotiated process of research.

Beyond these first two chapters of the book, which provided an insight into how and why the work underlying it took place, I set out a series of five more substantive and inter-linked chapters. In chapter 3, I explored the obviously significant and recurring experiential themes of panic and consumption, in relation to the early existentialist theorizing of Kierkegaard. This exploration illuminated the connections between these strands in terms of a certain closeness between the nature of agoraphobic panic (often experienced in and/or related to consumer spaces) and existential *Angst*. This latter form of unnerving emotion is conceptualized by Kierkegaard as the sense of a fundamental threat to our very *Being*, an awareness of the *abyss* or nothingness that necessarily underlies our existence in the world. In the (rather gloomy) light of this existentialist worldview, the overwhelming attack of agoraphobic panic can, I argued, be seen as a sudden crystallization of *Angst*, experienced as the near dissolution of one's sense of self. That is to say, panic *questions* the very grounds of one's existence and identity, the basic sense that one *is* something, rather than nothing.

Following this chapter, the focus initially shifted to consider what could be described as the experiential and conceptual opposite of consumer spaces for the agoraphobic; that is, the soothing and (mostly) panic free, safe space of the home. Here, behind closed doors, one is generally free from the objectifying 'look' of others, and the constitution and maintenance of self-identity is an almost effortless 'performance', executed on one's own territory and terms. In stark contrast, the antithetical pathoscapes of social space (epitomized by the mall discussed in chapter 3) present a deeply disturbing challenge to the agoraphobic's ability to project an acceptable (read 'normal') 'public' image. Social space is perceived as a highly pressured and pressurizing environment, one that intrudes on the boundaries of fragile selves and renders the task of putting on an appropriate, protective 'face' in order to 'blend in' virtually impossible.

Clearly, unless the agoraphobic subject can establish some way of managing these social anxieties, very little progress beyond their home and associated safe space is likely to be made. This chapter suggests that, in response to the insights Sartre provides into this socio-spatial problematic, via his account of the objectifying 'look', we might turn to Goffman's analysis in search of potential means for resolution. Goffman's perceptive account of the practical and largely unnoticed minutiae of psychological survival in social spaces could, I argued, prove invaluable for agoraphobic therapies. The issue of therapy was taken up in more depth in the final chapter, where I argued that whatever form or course is devised for individual agoraphobics, it must take account of the power of others to influence the shape and texture of social space. In consequence, it must be able to offer some means of self-protection to the subject threatened by the emotional projections, spatial constructions, and the 'look' of others. Chapter 4's appropriation of Sartre is thus more socially grounded and 'down to earth' than its rather esoteric predecessor, but still lies some way short of the fully empathetic, feminist and phenomenological account which the book aimed to develop.

In evident contrast with what has gone before, chapter 5 turned to explore the situation and experience of just one respondent in particular, phenomenological depth. Merleau-Ponty's approach was used here to highlight the peculiar and inescapable embodiment of agoraphobia, and to bring a live, panicked dimension to the rather bloodless and distant approaches of early existentialist theorists. It thus shifted the focus from the nature of 'Being' or existence as an abstract concept, to the fleshy realities of being-involved with the mundane practices, and so also anxieties, of everyday life. Perception is obviously key to Merleau-Ponty's worldview, and his approach emphasizes the importance of our complex sensory involvement with our environs for understandings and experience of agoraphobia. One's sense of self-identity is shown to be mediated by multiple (i.e. not only visual) sensations, and significantly, to be mediated in a manner that transcends dualistic distinctions between one's mind and one's body. Merleau-Ponty's approach thus moves away from the Cartesian subject to enable a focus on mutually constitutive relations and fluid boundaries between (embodied) selves and 'surrounding' spaces.

The question(ing) of boundaries is an ever present theme in this book, and one that increasingly came to prominence as the phenomenological discussion proceeded through chapter 6. Here I explored possible links or 'synergies' between contemporary experience of (insecurely bounded) womanhood, its exaggerated expression in pregnancy, and the embodiment of agoraphobia. This penultimate chapter once again underlined the importance of treating agoraphobia (in the theoretical as well as conventional sense of the term) in its embodied, socio-spatial contexts. It drew earlier Sartrean insights regarding objectification towards a specifically gendered experience of bodily boundaries, combined with a phenomenological reading of how these boundaries are troubled and transformed. This chapter concluded that the contemporary socio-cultural climate in which western women live works to undermine the stability of our embodied senses of self, via judgemental and degrading practices of emotionalization, objectification, exclusion and infantilization. The chapter argued that agoraphobia might be thought of as a boundary crisis arising from these masculinist cultural conditions.

Such boundary explorations lead us on to issues of treatment and thus back, in the final chapter, to the self-help groups with which we began. We returned here to look in critical detail at some therapeutic options currently available to the subject in (boundary) crisis, and to their own assessment of the potential value of these resources. This chapter emphasized the importance of listening to sufferers' own descriptions of their embodied experience, given the phenomenal distance between the version of panic they themselves present, and that predicated of the Cartesian (masculine) subject who frequently seems to star in self-help resources. Chapter 7 suggested that agoraphobic subjects may well benefit from the usage of therapies based on feminist models of selves, models that allow for differently felt and disparately constituted boundaries.

It has hopefully, then, become increasingly clear as this book advanced that its intentions were to explore and open up, never 'fix', define or delimit, the experience and meanings of agoraphobia for the selves, and also societies, who are affected by its presence. As we moved through the 'minor theories' (Katz 1996) and stories of each chapter, the book attempted to reveal an increasingly nuanced, individualized and therapeutically inclined analysis of the potentially problematic, socially and spatially constituted nature of subjectivity. The insights brought to light are thus multi-perspectival but I hope that the particular 'take' on agoraphobic experience presented here has yielded valuable clues into the nature and roots of our inherently anxious (because always precarious) existence. Such philosophical inquiries into the nature of *Being* are not, I have argued, as esoteric and removed from everyday life as the terminology might suggest. Being is always already embodied and placed; to *be* is to be situated, and the *context* of day-to-day living is absolutely key. The increasingly intimate and intricate reading of the data generated by this project thus particularizes questions of women's existence, and does so in specific relation to embodiment in and by contemporary culture. This fully grounded approach involved analysis of the social construction – and crucially, resistance and disruption – of the spaces and places in which women move and are *moved*. As this statement implies, the relational impact between people and places is never simply unidirectional, and appreciation of the complex interplays *between* has implications for feminist treatment (again, theoretically and conventionally) of situated selves.

With regard to the emotional influence and 'power' of place, the book clearly demonstrates that we can be moved by, for example shopping malls, to feel fear, frustration and panic. However, respondents have shown that these self-same places can also move us to feel anger, determination, and sometimes, even hope. Thus, such emotional states are not always 'steady' or predictable, and can (be made to?) change over time. Emotions such as anxiety are interactive, rather than simply responsive. They come into being at the shifting interface *between* people and their environments, a dynamic and mutually constitutive relation that means we can at least partially influence the nature of social space, and in turn, the nature of its influence on our selves. This conclusion leads me to argue that a particular embodiment of feminism – of feminist (emotional) ideas and (bodily) attitudes – could inform and help *perform* particular kinds of space (Rose 1999), and thus enable more comfortable and even, almost unthinkably, *enjoyable* spatial engagements. Here, one might learn from Susan Bordo's personal account of the shift from feeling 'substanceless' in social space to having a sense of 'definition' – metaphors strongly evocative of the absence and presence of boundaries – and experiencing an *exciting* rather than incapacitating 'charge' at the thought of leaving home (in Bordo et al 1998; see also chapter 3). Negative and panicky 'reactions' are not set in stone, and feminist insights into their incipient existence at the boundaries of our lives can help us work towards their transformation.

I thus want to close this book by suggesting that we can usefully draw on Bordo's and other feminist accounts of re-formed self/space relations, and put such embodied musings into practice for the benefit of phobic and other uncomfortably bounded women. Feminist theory and therapy, I have argued, can help us deal with our embodiment in and objectification by contemporary masculinist society. Such large-scale problems inevitably have to be handled at the smaller scale of the 'geography closest in'. Feminism works at multiple levels, and can be brought to work im*mediately* through the bodies and boundaries of the individual, helping to strengthen her resistance to subversions of her self. As we have seen (chapter 4), exposure to the look of others – an objectifying, and clearly gendered phenomenon – heightens awareness of the precarious nature of Being, and of being next to *Nothing*. It powerfully evokes the frightening closeness of the substanceless abyss and in our defence, boundaries must be brought into Being. They must be *made* to lie between us, and this is a project that ought to be undertaken inter-subjectively. Given that boundaries are negotiated in relation, the individual alone need not be held responsible for their production and maintenance. In this context, feminist approaches to boundary 'therapy' could usefully consider the role and value of women's self-help, or more appropriately, 'mutual support' groups. There is, I would argue, much potentially valuable research to be conducted in this area, given the possible benefits for emotional health of exchanges between women with similar but shifting embodied experience of boundaries. It is perhaps in this supportive context that the possibilities for and effects of the social and spatial performance of 'boldness' (Koskela 1997), among other assertions of feminist self-hood and spatiality, could most productively be explored.

Finally, to recap, having identified and theorized the nature of agoraphobic, and by implication 'other' socio-spatial anxieties, in the phenomenological terms of boundary crises, this book offers a critique of an existing and predominant treatment model, and points towards possible alternatives and directions for future research. I have argued that, despite what may be deeply held and long-term, even *ingrained* anxieties in the face of social space, feminist informed therapeutic intervention might enable a degree of re-negotiation, and substantiation, of our felt sense of the bodily boundaries that distinguish between yet inter-connect selves with 'surrounding' spaces. As we have seen (particularly in chapters 5 and 6), the projection of boundaries is a bodily habit and attitude towards the world that needs to be *practised* and regularly exercised. Perhaps, then, by initially doing so in a supportive therapeutic environment we can (re-?) learn to intensify and strengthen our (perception of our) boundaries, and to thus fabricate a 'space for freedom' around our selves (Jaspers in Young-Bruehl 1981: xi). With an embodied sense of such peripheral protection we would be better placed to mediate our involvement with social space and be better protected against the projections and potential intrusions of its others.

Bibliography

Ackroyd, Stephen and Hughes, John A. (1992) *Data Collection in Context* (New York: Longman Inc.).

Adams, Marie (2000) 'Humour in the psychotherapeutic relationship – not to be trifled with', *Counselling*, April.

Allwood, Robin (1996) 'I have depression, don't I?': Discourses of help and self-help books', in Erica Burman (ed.) *Psychology, Discourse, Practice* (London: Taylor and Francis).

American Psychiatric Association (1994) *DSM IV: Diagnostic and statistical manual of mental disorders* (4th ed.), (Washington, DC).

Arendt, Hannah (1958) *The Human Condition* (Chicago: Chicago University Press).

Astbury, Jill (1996) *Crazy for You: The Making of Women's Madness* (Oxford: Oxford University Press).

Bailey, Cathy, White, Catherine and Pain, Rachel (1999) 'Evaluating Qualitative Research: Dealing with the Tension Between 'Science' and 'Creativity'', *Area*, 31, 2, 169-183.

Bailey, Lucy (1999) Refracted Selves? A Study of Changes in Self-Identity in the Transition to Motherhood, *Sociology*, 33, 2, 335-352.

Baker, Lynn Rudder (1984) 'On the very idea of a form of life', *Inquiry*, 27, 277-89.

Baker, Roger (ed.) (1995) *Panic Disorder: Theory, Research and Therapy* (Sussex: John Wiley & Sons Ltd.).

Bankey, Ruth (2001) La donna e mobile: Constructing the irrational woman, *Gender, Place and Culture: A Journal of Feminist Geography*, 8, 1, 37-54.

Bannan, John F. (1967) *The Philosophy of Merleau-Ponty* (New York: Harcourt, Brace and World Inc.).

Barlow, David H. and Cerny, Jerome A. (eds.) (1988) *Psychological Treatment of Panic* (New York: The Guilford Press).

Bartlett, Dean and Payne, Sheila (1997) 'Grounded Theory: Its Basis, Rationale and Procedures', in McKenzie, George, Powell, Jackie and Usher, Robin (eds.) *Understanding Social Research: Perspectives on Methodology and Practice* (London: The Falmer Press).

Battersby, Christine. (1998) *The Phenomenal Woman: Feminist Metaphysics and the Patterns of Identity* (Cambridge: Polity Press).

Bauman, Zygmunt (1988) 'Sociology and Postmodernity', *Sociological Review*, 36, 4, 790-813.

Bekker, Marrie H.J. (1996) 'Agoraphobia and Gender: A Review', *Clinical Psychology Review*, 16, 2, 129-146.

Bingley, Amanda (2002) 'Research Ethics in Practice', in Bondi et al. *Subjectivities, Knowledges and Feminist Geographies* (Oxford: Rowman and Littlefield).

Bondi, Liz (1992) 'Gender and dichotomy', *Progress in Human Geography*, 16, 1, 98-104.

Bondi, Liz (1999a) 'Stages on Journeys: Some Remarks about Human Geography and Psychotherapeutic Practice', *Professional Geographer*, 5, 1, 11-24.

Bondi, Liz (1999b) 'Spaces for learning', unpublished paper presented at the International Symposium on Knowledge, Education and Space, Heidleberg.

Bondi, Liz and Burman, Erica (2001) 'Women and Mental Health: A Feminist Review', *Feminist Review*, 68, 6-33.

Bondi, Liz and Domosh, Mona (1998) 'On the contours of public space: a tale of three women', *Antipode*, 30, 270-289.

Bondi, Liz, Avis, Hannah, Bankey, Ruth, Bingley, Amanda, Davidson, Joyce, Duffy, Rosaleen, Einagel, Victoria Ingrid, Green, Anja-Maaike, Johnston, Lynda, Lilley, Susan, Listerborn, Carina, McEwan, Shonagh, Marshy, Mona, O'Connor, Niamh, Rose, Gillian and Vivat, Bella (2002) *Subjectivities, Knowledges and Feminist Geographies: The Subjects and Ethics of Social Research* (Oxford: Rowman and Littlefield).

Bordo, Susan (1988) 'Anorexia Nervosa: Psychopathology as the Crystallization of Culture', in Irene Diamond & Lee Quinby (eds.) *Feminism and Foucault: Reflections on Resistance* (Boston: Northeastern University Press).

Bordo, Susan (1993) *Unbearable Weight: Feminism, Western Culture, and The Body* (Berkeley: University of California Press).

Bordo, Susan, Klein, Binnie and Silverman, Marilyn K. (1998) 'Missing Kitchens', in Heidi Nast and Steve Pile (eds.) *Places Through the Body* (London and New York: Routledge).

Bowlby, John (1973) *Separation: Anxiety and Anger* (London: Hogarth Press).

Brehony, K A (1983) 'Women and Agoraphobia: A Case for the Etiological Significance of the Feminine Sex-Role Stereotype', in Violet Franks and Esther Rothblum (eds.) *The Stereotyping of Women: Its Effects on Mental Health* (New York: Springer Publishing Company).

Britten, Nicky (2000) 'Qualitative Interviews in Health Care Research', in Catherine Pope and Nicholas Mays (eds.) *Qualitative Research in Health Care* (London: BMJ).

Brown, Gillian (1987) 'The Empire of Agoraphobia', *Representations*, 20, 134-157.

Brown, Laura (1995) *Subversive Dialogues: Theory in Feminist Therapy* (New York Basic Books).

Brown, Laura S. and Root, Maria P. (1996) *Diversity and Complexity in Feminist Therapy* (New York: Haworth Press).

Bruch, Hilde (1979) *The Golden Cage: The Enigma of Anorexia Nervosa* (New York: Vintage).

Burgess, R. C. (1984) *In the Field: An Introduction to Field Research* (London and New York: Routledge).

Burgin, Victor (1996) *In/Different Spaces: Place & Memory in Visual Culture* (Berkeley, Los Angeles and London: University of California Press).

Burman, Erica (1998) *Deconstructing Feminist Psychology* (London: Sage).

Busfield, Joan (1984) 'Is mental illness a female malady? Men, women and madness in nineteenth century England', *Sociology*, 28, 259-277.

Busfield, Joan (1996) *Men, Women and Madness: Understanding Gender and Mental Disorder* (London: MacMillan Press).

Butler, Judith (1990) *Gender Trouble: Feminism and the Subversion of Identity* (London and New York: Routledge).

Butler, Judith (1993) *Bodies That Matter: On the Discursive Limits of 'Sex'* (London and New York: Routledge).

Butler, Ruth and Hester Parr (1999) *Mind and Body Spaces: Geographies of Illness, Impairment and Disability* (London and New York: Routledge).

Caillois, Roger (1984) 'Mimicry and Legendary Psychasthenia', *October 31*, Winter, 17-32.

Campbell, Colin (1997) 'Shopping, Pleasure and the Sex War', in Falk, Pasi and Campbell, Colin (eds.) (1997) *The Shopping Experience* (London: Sage Publications).

Capps, Lisa and Ochs, Elinor (1995) *Constructing Panic: The Discourse of Agoraphobia* (Cambridge and London: Harvard University Press).

Chambless, Dianne L. (1982) 'Characteristics of Agoraphobia' in Chambless, Dianne L. and Goldstein, Alan J. (eds.) *Agoraphobia: Multiple Perspectives on Theory and Treatment* (New York: John Wiley & Sons).

Chambless, Dianne L. & Goldstein, Alan J. (1980) 'Anxieties, Agoraphobia and Hysteria', in Annette M. Brodsky and Rachel Hare-Mustin (eds.) *Women in Psychotherapy: An Assessment of Research and Practice* (New York: The Guilford Press).

Chambless, Dianne L. & Goldstein, Alan J. (eds) (1982) *Agoraphobia: Multiple Perspectives on Theory and Treatment* (New York: John Wiley & Sons).

Chambless, Dianne L. and Mason, Jeanne. (1986) 'Sex, sex-role stereotyping and agoraphobia', *Behaviour, Research and Therapy*, 24, 2, 231-235.

Chaplin, Jocelyn (1997) *Feminist Counselling in Action* (London and Thousand Oaks: Sage).

Chernin, Kim (1981) *The Obsession: Reflections on the Tyranny of Slenderness*, (New York: Harper and Row).

Chodorow, Nancy (1978) *The Reproduction of Mothering: Psychoanalysis and the Sociology of Gender* (Berkeley: University of California Press).

Chodorow, Nancy (1994) *Femininities, Masculinities, Sexualities: Freud and Beyond* (London: Free Association Books).

Clum, G. A. and Knowles, S. L. (1991) 'Why do some people with panic disorder become avoidant?: A review', *Clinical Psychology Review*, 11, 295-313.

Cooper, David E. (1990) *Existentialism* (Oxford: Blackwell).

Corbin, J. and Strauss, A. (1990) 'Grounded Theory Research: Procedures, Canons and Evaluative Criteria', *Qualitative Sociology*, 13, 1, 3-21.

Crawford, Margaret (1994) 'The World In A Shopping Mall', in Michael Sorkin (ed.) *Variations On A Theme Park: The New American City and the End of Public Space* (New York: Hill and Wang).

Csordas, Thomas J. (ed.) (1994) *Embodiment and Experience: The Existential Ground of Culture and Self* (Cambridge: Cambridge University Press).

da Costa Meyer, Esther (1996) 'La donna e mobile: Agoraphobia, women, and urban space', in Diana Agrest, Patricia Conway and Leslie Kanes Weisman (eds.) *The Sex of Architecture* (New York: Harry N. Abrams).

Davidson, Joyce and Smith, Mick (1999) 'Wittgenstein and Irigaray: Gender and Philosophy in a Language (Game) of Difference', *Hypatia*, 14, 2, 72-96.

Davis, Madeleine and Wallbridge, David (1992) *Boundary and Space: An Introduction to the Work of D. W. Winnicott* (New York: Bruner/Mazel Publishers).

de Swaan, Abram (1981) 'The Politics of Agoraphobia: On Changes in Emotional and Relational Management', *Theory and Society*, 10, 359-385.

Deutsche, Rosalyn (1990) 'On men in space', *Artforum*, 21-23.

Devault, Marjorie (1990) 'Talking and listening from women's standpoint: feminist strategies for interviewing and analysis', *Social Problems*, 37, 1, 96-116.

Dinnerstein, Dorothy (1978) *The Mermaid and the Minotaur: Sexual Arrangements and Human Malaise* (New York: Harper and Row).

Domosh, Mona (1998) 'Geography and gender: home, again?', *Progress in Human Geography*, 22, 2, 276-282.

Dorn, Michael (1998) 'Beyond nomadism: the travel narratives of a "cripple"', in Nast, Heidi J. and Pile, Steve (eds.) *Places Through the Body* (London and New York: Routledge).

Dowling, Robyn (1998) 'Suburban stories, gendered lives: thinking through difference', in Ruth Fincher and Jane Jacobs (eds.) *Cities of Difference* (London and New York: The Guilford Press).

Duley, Bridget (1997) 'Authoring New Zealand: Media Representations Of National Identity', Unpublished Thesis, University of Waikato.

Dunant, Sarah and Porter, Roy (eds.) (1996) *The Age of Anxiety* (London: Virago).

Duncan, Nancy (1996) 'Renegotiating gender and sexuality in public and private spaces', in Nancy Duncan (ed.) *BodySpace* (London and New York: Routledge).

Dyck, Isabel (1995) 'Hidden geographies: the changing lifeworlds of women with multiple sclerosis', *Social Science and Medicine*, 40, 307-320.

Dyck, Isabel (1993) 'Ethnography: A Feminist Method?', *The Canadian Geographer*, 37, 1, 52-57.

Edie, James M. (1987) *Edmund Husserl's Phenomenology: A Critical Commentary* (Bloomington, Indianapolis: Indiana University Press).

Edwards, Sandra and Uhlenhuth, E.H. (1998) 'Panic Disorder and Agoraphobia: A Sufferer's Perspective', *Journal of Affective Disorders*, 50, 65-74.

Eisenbaum, Luise and Orbach, Susie (1992) *Understanding Women* (London: Penguin Books).

England, Kim V.L. (1994) 'Getting personal: reflexivity, positionality, and feminist research', *The Professional Geographer*, 46, 80-89.

Evernden, Neil (1985) *The Natural Alien: Humankind and Environment* (Toronto: University of Toronto Press).

Fairclough, Norman (1992) *Discourse and Social Change* (Cambridge: Polity Press).

Fairclough, Norman (1995) *Media Discourse* (New York: Edward Arnold).

Falk, Pasi and Campbell, Colin (eds.) (1997) *The Shopping Experience* (London: Sage Publications).

Featherstone, Mike (1996) *Consumer Culture & Postmodernism* (London: Sage Publications).

Flax, Jane (1990) *Thinking Fragments: Psychoanalysis, Feminism and Post-modernism in the Contemporary West* (Berkeley: University of California Press).

Fodor, Iris Goldstein (1974) The phobic syndrome in women: Implications for treatment, in: Violet Franks & Vasanti Burtle (eds.) *Women in Therapy: New Psychotherapies for a Changing Society* (New York: Bruner Mazel).

Fogelin, Robert J. (1976) *Wittgenstein* (London: Routledge & Kegan Paul).

Foucault, Michel (1984) 'The Order of Discourse', in M. Shapiro (ed.) *Language and Politics: Readings in Social and Political Theory*, trans. I. McLeod (Oxford: Basil Blackwell).

Freely, Maureen (2000) 'Cherie's Choice', *The Guardian*, May 19, G2, 4.

Freud, Sigmund (1960) *Jokes and their Relation to the Unconscious* (Harmondsworth: Penguin).

Freund, Peter (1998) 'Social Performances and their Discontents: The Biopsychosocial Aspects of Dramaturgical Stress', in Gillian Bendelow and Simon J. Williams (eds.) *Emotions in Social Life* (London and New York: Routledge).

Gabriel, Yiannis and Lang, Tim (1995) *The Unmanageable Consumer: Contemporary Consumption and its Fragmentations* (London: Sage Publications).

Garbowsky, Maryanne M. (1989) *The House Without the Door: A Study of Emily Dickinson and the Illness of Agoraphobia* (London and Toronto: Associated University Presses).

Gardner, Carol Brooks (1994) 'Out of Place: Gender, Public Places, and Situational Disadvantage', in R. Friedland and D. Boden (eds.), *NowHere: Space, Time and Modernity* (Berkeley: University of California Press).

Gartner, Alan and Reissman, Frank (eds.) (1984) *The Self-Help Revolution* (New York: Human Sciences Press Inc.).

Gelder, Michael; Gath, Dennis; Mayou, Richard and Cowen, Philip (eds.) (1996) *Oxford Textbook of Psychiatry* [3rd ed.] (Oxford: Oxford University Press).

Giddens, Anthony (1997) *Modernity and Self-Identity* (Cambridge: Polity Press).

Gilbert, Melissa (1994) 'The politics of location: doing feminist research at "home"', *The Professional Geographer*, 46, 1, 90-96.

Gilligan, Carol (1982) *In a Different Voice: Psychoanalytic Theory and Women's Development* (Cambridge, Massachusetts: Harvard University Press).

Glaser, B. G. and Strauss, A. L. (1967) *The Discovery of Grounded Theory: Strategies for Qualitative Research* (London: Weidenfield and Nicolson).

Goffman, Erving (1963) *Behavior in Public Places: Notes on the Social Organisation of Gatherings* (London: Collier Macmillan Ltd.).

Goffman, Erving (1967) *Interaction Ritual* (Harmondsworth: Penguin Books Ltd).

Goffman, Erving (1969) *Where The Action Is* (London: Allen Lane, The Penguin Press).

Goffman, Erving (1971) *Relations in Public: Microstudies of the Public Order* (Harmondsworth: Penguin Books Ltd).

Goffman, Erving (1980) *The Presentation of Self in Everyday Life* (Harmondsworth: Penguin Books Ltd).

Goffman, Erving (1990 [1963]) *Stigma: Notes on the Management of Spoiled Identity* (Harmondsworth: Penguin Books Ltd).

Goss, Jon (1992) 'Modernity and post-modernity in the retail landscape', in Kay Anderson and Fay Gale (eds.) *Inventing Places: Studies in Cultural Geography* (Cheshire: Longman).

Goss, Jon (1993) 'The "Magic of the Mall": An Analysis of Form, Function, and Meaning in the Contemporary Retail Built Environment', *Annals of the Association of American Geographers*, 83, 1, 18-47.

Gournay, Kevin (ed.) (1989) 'Introduction: The Nature of Agoraphobia and Contemporary Issues', *Agoraphobia: Current Perspectives on Theory and Treatment* (London and New York: Routledge).

Grieder, Alfons (1999) 'Phenomenology and Psychotherapy', *Changes* 17, 3, 178–187.

Gregson, Nicky (1995) 'And now its all consumption?', *Progress in Human Geography* 19, 1, 135-141.

Gregson, Nicky and Gillian Rose (2000) 'Taking Butler Elsewhere: Performativities, Spatialities and Subjectivities', *Environment and Planning D: Society and Space* 18: 433-452.

Grosz, Elizabeth (1994) *Volatile Bodies: Towards a Corporeal Feminism* (Bloomington and Indianapolis: Indiana University Press).

Hallam, R. (1992) *Counselling For Anxiety Problems* (London: Sage Publications Ltd.).

Hammond, M.A., Howarth, Jane M. and Keat, R.N. (1991) *Understanding Phenomenology* (Massachusetts: Basil Blackwell).

Hannay, Alastair (1982) *Kierkegaard* (London: Routledge & Kegan Paul).

Hannay, Alastair (1989) Introduction to Kierkegaard, Søren *The Sickness Unto Death: A Christian Psychological Exposition for Edification and Awakening* [trans. Alastair Hannay] (London: Penguin Books).

Heidegger, Martin (1988) *Being and Time* (Oxford: Blackwell).

Heidegger, Martin. (1993) 'Building, Dwelling, Thinking' in *Basic Writings* (London: Routledge).

Hekman, Susan (1990) *Gender and Knowledge: Elements of a Post-modern Feminism* (Cambridge: Polity Press).

Hopkins, J.S.P. (1990) 'West Edmonton Mall: Landscapes of myths and elsewhereness', *Canadian Geographer* 34, 1, 2-17.

Horkheimer, Max and Adorno, Theodore (1972) *The Dialectic of Enlightenment* (New York: Herder & Herder).

Irigaray, Luce (1985) *Speculum of the Other Woman* (Ithaca New York: Cornell University Press).

Irigaray, Luce (1985) 'The Mechanics of Fluids' in *This Sex Which Is Not One* (Ithaca, New York: Cornell University Press).

Jones, Colin and Porter, Roy. (eds.) (1998) *Reassessing Foucault: Power, Medicine and the Body* (London and New York: Routledge).

Katz, Alfred (1984) 'Self-Help Groups: An International Perspective', in Gartner and Reissman (eds.) *The Self-Help Revolution* (New York: Human Sciences Press Inc.).

Katz, Cindi (1994) 'Playing the field: questions of fieldwork in geography', *The Professional Geographer*, 46, 1, 67-72.

Katz, Cindi (1996) 'Towards Minor Theory', *Environment and Planning D; Society and Space*, 14, 487-499.

Kearns, Robin A. (1997) 'Narrative and metaphor in health geographies', *Progress in Human Geography* 21, 2, 269-277.

Kierkegaard, Søren (1980) *The Concept of Anxiety: A Simple Psychologically Orienting Deliberation on the Dogmatic Issue of Hereditary Sin* [trans. Reidar Thomte in collaboration with Albert B. Anderson] (Princeton and Chichester: Princeton University Press).

Kierkegaard, Søren (1985) *Fear and Trembling: Dialectical Lyric* [trans. Alastair Hannay] (London: Penguin Books).

Kierkegaard, Søren (1989) *The Sickness Unto Death: A Christian Psychological Exposition for Edification and Awakening* [trans. Alastair Hannay] (London: Penguin Books).

Kirby, Kathleen M. (1996) *Indifferent Boundaries: Spatial Concepts of Human Subjectivity* (New York and London: The Guilford Press).

Kobayashi, Audrey (1994) 'Coloring the field: "race," and the politics of fieldwork', *The Professional Geographer*, 46, 1, 73-80.

Koskela, Hille (1997) '"Bold Walk and Breakings": women's spatial confidence versus fear of violence', *Gender, Place and Culture*, 4, 3, 301-320.

Kowinski, W. S. (1985) *The Malling of America: An Inside Look at the Great Consumer Paradise* (New York: William Morrow).

Krell, David Farrell (1993) in Martin Heidegger, *Basic Writings* (London: Routledge).

Kristeva, Julia (1982) *Powers of Horror: An Essay on Abjection* (New York: Columbia University Press).

Kroker, Arthur, Kroker, Marilouise and Cook, David (1989) *Panic Encyclopedia: The Definitive Guide to the Postmodern Scene* (London: MacMillan Education Ltd.).

Kvale, Steinar (1996) *InterViews: An Introduction to Qualitative Research Interviewing* (London: Sage).

Langman, Lauren (1994) 'Neon Cages: Shopping for Subjectivity', in Rob Shields (ed.) *Lifestyle Shopping: The Subject of Consumption* (London & New York: Routledge).

Law, S. N. (1975) *Inspired Freedom: Agoraphobia - A Battle Won* (Regency Press: London and New York).

Leder, Drew (1990) *The Absent Body* (Chicago: The University of Chicago Press).

Lefebvre, Henri (1994) *The Production of Space* (Oxford: Blackwell).

Lemma, Alessandra (2000) *Humour on the Couch: Exploring Humour in Psychotherapy and in Everyday Life* (London: Whurr Publishers Ltd.).

Levy, A. (1977) 'Devant et derriere soi', *Nouvelle revue de psychanalyse*, 15, Spring, 93-94.

Lilley, Susan M. (1998) *The Socio-technical Production of GIS Knowledges: The Discursive Construction of Bodies and Machines at Scottish Natural Heritage*, unpublished PhD thesis, University of Edinburgh.

Longhurst, Robyn (1996) 'Refocusing groups', *Area*, 28, 2, 143-149.

Longhurst, Robyn (1997) '"Going nuts": re-presenting pregnant women', *New Zealand Geographer*, 53, 2, 34-39.

Longhurst, Robyn (1998) '(Re)presenting shopping centres and bodies: questions of pregnancy', in: Rosa Ainley (ed.) *New Frontiers of Space, Bodies, Gender* (London and New York: Routledge).

Longhurst, Robyn (1999) Attempts to impose limits: the disorderly bodies of pregnant women, in: Elizabeth Kenworthy Teather (ed.) *Embodied Geographies: Spaces, Bodies and Rites of Passage* (London and New York: Routledge).

Longhurst, Robyn (2000a) *Bodies: Exploring Fluid Boundaries* (London and New York: Routledge).

Longhurst, Robyn (2000b) 'Corporeographies of Pregnancy: "Bikini Babes"', *Environment and Planning D: Society and Space*, 18, 4, 453-472.

Lorraine, Tamsin. (1999) *Irigaray and Deleuze: Experiments in Visceral Philosophy* (New York: Cornell University Press).

Lowe, M. and Crewe, L. (1991) 'Lollipop jobs for pin money? Retail employment explored', *Area*, 23, 344-347.

Lupton, Deborah (1994) *Medicine as Culture: Illness, Disease and the Body in Western Societies* (London: Sage).

MacPherson, C. B. (1979) *The Political Theory of Possessive Individualism: Hobbes to Locke* (Oxford: Oxford University Press).

Macquarrie, John (1971) *Existentialism* (London: Hutchinson & Co. Ltd.).

Marks, Isaac M. (1975) [1969] *Fears and Phobias* (London: William Heinemann Medical Books).

Marks, Isaac M. (1980) *Living With Fear: Understanding and Coping with Anxiety* (New York: McGraw Hill).

Marks, Isaac M. (1987a) *Fears, Phobias and Rituals* (New York and Oxford: Oxford University Press).

Marks, Isaac M. (1987b) 'Agoraphobia, panic disorder and related conditions in the DSM IIIR and ICD 10', *Journal of Psychopharmacology*, 1, 6-12.

Marks, Isaac (1995) 'Agoraphobia and Panic Disorder', in Baker, Roger (ed.) *Panic Disorder: Theory, Research and Therapy* (Sussex: John Wiley & Sons Ltd.).

Martin, P. (1997) *The Sickening Mind* (London: Harper Collins).

Massey, Doreen (1992) 'Politics and Space/Time', *New Left Review*, 196, 65-84.

Massey, Doreen (1993) 'Power Geometry and a Progressive Sense of Place', in J. Bird, B. Curtis, G. Robertson and L. Tickner (eds.) *Mapping the Futures: Local Culture, Global Change* (London and New York: Routledge).

Mathews, Andrew M., Gelder, Michael G., Johnston, Derek W. (1981) *Agoraphobia: Nature and Treatment* (London and New York: Tavistock Publications).

Maynard, Mary (1994) 'Methods, Practice and Epistemology: The Debate about Feminism and Research', in Maynard, Mary and Purvis, June (eds.) (1994) *Researching Women's Lives from a Feminist Perspective* (London: Taylor & Francis).

Maynard, Mary and Purvis, June (eds.) (1994) *Researching Women's Lives from a Feminist Perspective* (London: Taylor & Francis).

McCracken, G. (1988) *The Long Interview* (London: Sage).

McDowell, Linda (1992) 'Doing gender: feminism, feminists and research methods in human geography', *Transactions, Institute of British Geographers*, 17, 399-416.

McDowell, Linda (1996) 'Spatializing Feminism: Geographic Perspectives', in *BodySpace*, Nancy Duncan (ed.) (London and New York: Routledge).

McDowell, Linda (1999) *Gender, Place, Identity* (Cambridge: Polity Press).

McKenzie, George, Powell, Jackie and Usher, Robin (eds.) (1997) *Understanding Social Research: Perspectives on Methodology and Practice* (London: The Falmer Press).

McLeod, Eileen (1994) *Women's Experience of Feminist Therapy and Counselling* (Milton Keynes: Open University Press).

McNally, Richard J. (1994) *Panic Disorder: A Critical Analysis* (New York and London: The Guilford Press).

Melville, Joy (1979) *Phobias* (London: Unwin Paperbacks).

Merleau-Ponty, Maurice (1962) *Phenomenology of Perception* [trans. Colin Smith], (London: Routledge & Kegan Paul).

Mies, Maria (1983) 'Towards a Methodology for Feminist Research', in G. Bowles and R.D. Klein (eds.) *Theories for Women's Studies* (London: Routledge & Kegan Paul).

Miller, Daniel (1997) 'Could Shopping Ever Really Matter', in Falk, Pasi and Campbell, Colin (eds.) *The Shopping Experience* (London: Sage Publications).

Moi, Toril (1999) *What is a Woman? And Other Essays* (Oxford: Oxford University Press).

Moore, S. (1991) *Looking for Trouble: On Shopping, Gender and the Cinema* (London: Serpent's Tail).

Moran, Dermot (2000) *Phenomenology* (London & New York: Routledge).

Morse, J. M. (1992) *Qualitative Health Research* (London: Sage).

Moss, Pamela and Dyck, Isabel (1999) 'Body, corporeal space and legitimacy. Chronic illness: women diagnosed with ME', *Antipode*, 31, 4, 372-397.

Moss, Pamela, Eyles, John, Dyck, Isabel and Rose, Damaris (1993) 'Focus: Feminism As Method', *The Canadian Geographer*, 37, 1, 48-61.

Mumford, Lewis (1991) *The City in History: Its Origins, its Transformations and its Prospects* (London: Penguin).

Murphy, Julien S. (1989) 'The Look in Sartre and Rich', in Jeffner Allen and Iris Marion Young (eds.) *The Thinking Muse: Feminism and Modern French Philosophy* (Bloomington and Indianapolis: Indiana University Press).

Murphy, Julien S. (ed.) (1999) *Feminist Interpretations of Jean-Paul Sartre* (Pennsylvania: Pennsylvania State University Press).

Nast, Heidi J. (1994) 'Opening Remarks on "Women in the Field: Critical Feminist Methodologies and Theoretical Perspectives"', *Professional Geographer*, 46, 1, 54-66.

Nast, Heidi J. and Pile, Steve (eds.) (1998) *Places Through the Body* (London and New York: Routledge).

Orr, Jackie (1990) 'Theory on the Market: Panic, Incorporating', *Social Problems*, 7, 4, 460-484.

Orwell, George (1968) *The Collected Essays, Journalism and Letters of George Orwell. Vol. 3 As I Please*, S. Orwell and I. Angus (eds.) (London: Secker and Warburg).

Palmer, Jerry (1994) *Taking Humour Seriously* (London and New York: Routledge).

Park, Deborah C., Radford, John P. and Vickers, Michael H. (1998) 'Disability Studies in Human Geography', *Progress in Human Geography*, 22, 2, 208-233.

Parker, Iain (1992) *Discourse Dynamics: Critical Analysis for Social and Individual Psychology* (London and New York: Routledge).

Parr, Hester (1998) 'The politics of methodology in "post-medical geography": mental health research and the interview', *Health and Place*, 4, 341-353.

Parr, Hester (1999) 'Delusional Geographies: The Experiential Worlds of People During Madness/Illness', *Environment and Planning D: Society and Space*, 17, 673-690.

Parr, Hester (2000) 'Interpreting the Hidden Social Geographies of Mental Health: Ethnographies of Inclusion and Exclusion in Semi-Institutional Places', *Health and Place*, 6, 225-237.

Plumwood, Val. (1993) *Feminism and the Mastery of Nature* (London and New York: Routledge).

Pope, C., Ziebland, S. and Mays, N. (2000) 'Analysing qualitative data' in Catherine Pope and Nicholas Mays (eds.) *Qualitative Research in Health Care* (London: BMJ).

Powell, Chris (1988) 'A phenomenological analysis of humour in society', in Chris Powell and George E.C. Paton (eds.) *Humour in Society: Resistance and Control* (London: Macmillan).

Prior, Pauline M. (1999) *Gender and Mental Health* (London: MacMillan Press).

Probyn, Elspeth (1995) 'Lesbians in space. Gender, sex and the structure of missing', *Gender, Place and Culture*, 2, 1, 77-84.

Quinodoz, Danielle (1997) *Emotional Vertigo: Between Anxiety and Pleasure* [trans. Arnold Pomerans] (London and New York: Routledge).

Rabil, Albert (1967) *Merleau-Ponty: Existentialist of the Social World* (New York and London: Columbia University Press).

Rachman, S. (1998) *Anxiety* (Sussex: Psychology Press Ltd.).

Rapping, Elayne (1996) *The Culture of Recovery: Making Sense of the Self-Help Movement in Women's Lives* (Boston: Beacon Press).

Reeves, Joy B. (1986) 'Toward a Sociology of Agoraphobia', *Free Inquiry in Creative Sociology*, 14, 2, 153-158.

Rich, Adrienne (1976) *Of Woman Born: Motherhood as Experience and Institution* (New York: Bantam).

Rich, Adrienne (1986) *Blood, Bread and Poetry: Selected Prose 1979-1985* (New York: Norton).

Rose, Damaris (1993) 'On feminism, method and methods in human geography: an idiosyncratic overview', *The Canadian Geographer*, 37, 1, 57-61.

Rose, Gillian (1993) *Feminism and Geography: The Limits of Geographical Knowledge* (Polity Press: Cambridge).

Rose, Gillian (1999) 'Performing Space', in Doreen Massey, John Allen & Philip Sarre (eds.) *Human Geography Today* (Cambridge: Polity Press).

Rossman, Gretchen B. and Sharon F. Rallis (1998) *Learning in the Field: An Introduction to Social Research* (London: Sage).

Sartre, Jean-Paul (1965) *Nausea* (Harmondsworth: Penguin).

Sartre, Jean-Paul (1993) *Being and Nothingness* (London: Routledge).

Savage, Edna (1987) *Overcoming Agoraphobia* (Rochdale: Byron Press).

Schivelbusch, Wolfgang (1980) *The Railway Journey: Trains and Travel in the 19th Century* [trans. Anselm Hollo] (Oxford: Blackwell).

Searle, John R. (1988) *Intentionality: An Essay in the Philosophy of Mind* (Cambridge and New York: Cambridge University Press).

Shields, Rob (1989) 'Social spatialization and the built environment: West Edmonton Mall', *Environment and Planning D: Society and Space*, 7, 147-164.

Shields, Rob (ed.) (1994) *Lifestyle Shopping: The Subject of Consumption* (London and New York: Routledge).

Shimrat, Irit (1997) *Call Me Crazy: Stories From the Mad Movement* (Vancouver: Press Gang Publishers).

Showalter, Elaine (1987) *The Female Malady: Women, Madness and English Culture* (London: Virago).

Sibley, David (1995) *Geographies of Exclusion* (London and New York: Routledge).

Sims, Andrew (1983) *Neurosis In Society* (London and Basingstoke: MacMillan Press).

Skeggs, Beverley (1994) 'Situating the Product of Feminist Ethnography', in Mary Maynard and June Purvis (eds.) *Researching Women's Lives from a Feminist Perspective* (London: Taylor & Francis).

Smith, Dorothy (1987) *The Everyday World as Problematic: A Feminist Sociology* (Boston: Northeastern University Press).

Smith, Peter B. (ed.) (1980) *Small Groups and Personal Change* (London and New York: Methuen & Co. Ltd).

Soja, Edward (1989) *Postmodern Geographies: The Reassertion of Space in Critical Social Theory* (London: Verso).

Solomon, Michael (1994) *Consumer Behaviour: Buying, Having and Being* (Boston: Allyn & Bacon).

Spurling, Laurie. (1977) *Phenomenology and the Social World: The Philosophy of Merleau-Ponty and its Relation to the Social Sciences* (London, Henley and Boston: Routledge and Kegan Paul).

Straus, Erwin W. (1966) *Phenomenological Psychology* (London: Tavistock Publications Ltd.).

Strauss, Anselm (1987) *Qualitative Analysis for Social Scientists* (Cambridge: Cambridge University Press).

Strauss, A and Corbin J (1990) *Basics of Qualitative Research: Grounded Theory Procedures and Techniques* (London: Sage).

Teather, Elizabeth Kenworthy (1999) *Embodied Geographies: Spaces, Bodies and Rites of Passage* (London and New York: Routledge).

Thorpe, Geoffrey and Burns, Laurence (1983) *The Agoraphobic Syndrome: Behavioural Approaches to Evaluation and Treatment* (New York: John Wiley and Sons).

Tian, P.S., Wanstall, K. and Evans, L. (1990) 'Sex Differences in Panic Disorder with Agoraphobia', *Journal of Anxiety Disorders*, 4, 317-324.

Tooke, Jane (2000) 'Betweenness at Work', *Area*, 32.2, 217-223.

Tseelon, Efrat (1995) *The Masque of Femininity* (London: Sage).

Turner, Bryan (1987) *Medical Power and Social Knowledge* (London: Sage).

Unger, Rhoda K. (1998) *Resisting Gender: Twenty-five Years of Feminist Psychology* (London: Sage).

Ussher, Jane (1991) *Women's Madness: Misogyny or Mental Illness?* (Hemel Hempstead: Harvester Wheatsheaf).

Valentine, Gill (1996) '(Re)negotiating the "Heterosexual Street": Lesbian Productions of Space', in Nancy Duncan (ed.) *BodySpace* (London and New York: Routledge).

Valentine, Gill (1998) '"Sticks and stones may break my bones": a personal geography of harassment', *Antipode*, 30, 305-332.

Vidler, Anthony (1991) 'Agoraphobia: Spatial Estrangement in Simmel and Kracauer', *New German Critique*, 54, 31-45.

Vidler, Anthony (1992) *The Architectural Uncanny* (Cambridge, Massachusetts: MIT Press).

Vidler, Anthony (1993) 'Bodies in Space / Subjects in the City: Psychopathologies of Modern Urbanism', *Differences: A Journal of Feminist Cultural Studies*, 5, 3, 31-51.

Vidler, Anthony (2000) *Warped Space: Art, Architecture, and Anxiety in Modern Culture*, (Cambridge, Massachusetts: MIT Press).

Vines, Robyn (1987) *Agoraphobia: The fear of Panic* (London: Fontana Paperbacks).

Wiles, Rose (1994) '"I'm not fat I'm pregnant": The impact of pregnancy on fat women's body image', in Sue Wilkinson & Celia Kitzinger (eds.) *Women and Health: Feminist Perspectives* (London: Taylor & Francis).

Williamson, Janice (1992) 'I-less and Gaga in the West Edmonton Mall: Towards a Pedestrian Feminist Reading', in Dawn H. Currie and Valerie Raoul (eds.) *Anatomy of Gender: Women's Struggle for the Body* (Ottawa: Carleton University Press).

Williams, Simon J. and Bendelow, Gillian. (1998) *The Lived Body: Sociological Themes, Embodied Issues* (London and New York: Routledge).

Wittgenstein, Ludwig (1988) *Philosophical Investigations* (Oxford: Blackwell).

Yardley, Lucy (1994) *Vertigo and Dizziness* (London and New York: Routledge).

Yardley, Lucy (1997) *Material Discourses of Health and Illness* (London and New York: Routledge).

Young, Iris Marion (1990) 'The Ideal of Community and the Politics of Difference', in Linda J. Nicholson (ed.) *Feminism / Postmodernism* (London and New York: Routledge).

Young, Iris Marion (1990) *Throwing Like a Girl and Other Essays in Feminist Philosophy and Social Thought* (Bloomington and Indianapolis: Indiana University Press).

Young-Bruehl, Elisabeth (1981) *Freedom and Karl Jaspers Philosophy* (New Haven and London: Yale University Press).

Zaner, Richard M. (1964) *The Problem of Embodiment* (The Hague: Martinus Nijhoff).

Index

Printed and bound by CPI Group (UK) Ltd, Croydon, CR0 4YY

21/10/2024

01777088-0012